C. R. Bector, S. Chandra

Fuzzy Mathematical Programming and Fuzzy Matrix Games

Studies in Fuzziness and Soft Computing, Volume 169

Editor-in-chief
Prof. Janusz Kacprzyk
Systems Research Institute
Polish Academy of Sciences
ul. Newelska 6
01-447 Warsaw
Poland
E-mail: kacprzyk@ibspan.waw.pl

Further volumes of this series
can be found on our homepage:
springeronline.com

Vol. 153. J. Fulcher, L.C. Jain (Eds.)
Applied Intelligent Systems, 2004
ISBN 3-540-21153-5

Vol. 154. B. Liu
Uncertainty Theory, 2004
ISBN 3-540-21333-3

Vol. 155. G. Resconi, J.L. Jain
Intelligent Agents, 2004
ISBN 3-540-22003-8

Vol. 156. R. Tadeusiewicz, M.R. Ogiela
Medical Image Understanding Technology, 2004
ISBN 3-540-21985-4

Vol. 157. R.A. Aliev, F. Fazlollahi, R.R. Aliev
Soft Computing and its Applications in Business and Economics, 2004
ISBN 3-540-22138-7

Vol. 158. K.K. Dompere
Cost-Benefit Analysis and the Theory of Fuzzy Decisions – Identification and Measurement Theory, 2004
ISBN 3-540-22154-9

Vol. 159. E. Damiani, L.C. Jain, M. Madravia
Soft Computing in Software Engineering, 2004
ISBN 3-540-22030-5

Vol. 160. K.K. Dompere
Cost-Benefit Analysis and the Theory of Fuzzy Decisions – Fuzzy Value Theory, 2004
ISBN 3-540-22161-1

Vol. 161. N. Nedjah, L. de Macedo Mourelle (Eds.)
Evolvable Machines, 2005
ISBN 3-540-22905-1

Vol. 162. N. Ichalkaranje, R. Khosla, L.C. Jain
Design of Intelligent Multi-Agent Systems, 2005
ISBN 3-540-22913-2

Vol. 163. A. Ghosh, L.C. Jain (Eds.)
Evolutionary Computation in Data Mining, 2005
ISBN 3-540-22370-3

Vol. 164. M. Nikravesh, L.A. Zadeh, J. Kacprzyk (Eds.)
Soft Computing for Information Prodessing and Analysis, 2005
ISBN 3-540-22930-2

Vol. 165. A.F. Rocha, E. Massad, A. Pereira Jr.
The Brain: From Fuzzy Arithmetic to Quantum Computing, 2005
ISBN 3-540-21858-0

Vol. 166. W.E. Hart, N. Krasnogor, J.E. Smith (Eds.)
Recent Advances in Memetic Algorithms, 2005
ISBN 3-540-22904-3

Vol. 167. Y. Jin (Ed.)
Knowledge Incorporation in Evolutionary Computation, 2005
ISBN 3-540-22902-7

Vol. 168. Y. P. Tan, K. H. Yap, L. Wang (Eds.)
Intelligent Multimedia Processing with Soft Computing, 2004
ISBN 3-540-23053-X

C. R. Bector
Suresh Chandra

Fuzzy Mathematical Programming and Fuzzy Matrix Games

Dr. C. R. Bector
Department of Business Administration
Faculty of Management
Asper School of Business
University of Manitoba
Winnipeg, Manitoba
Canada, R3T 5V4

Dr. Suresh Chandra
Department of Mathematics
Indian Institute of Technology (I.I.T.)
Hauz Khas
New Delhi 110016
India

ISSN 1434-9922
ISBN 3-540-23729-1 Springer Berlin Heidelberg New York

Library of Congress Control Number: 2004116802

Springer is a part of Springer Science+Business Media
springeronline.com

Printed in Germany

Typesetting: data delivered by editor
Cover design: E. Kirchner, Springer, Heidelberg
Printed on acid free paper 62/3020/M - 5 4 3 2 1 0

DEDICATED
TO
Professor Dr. H. -J. Zimmermann
(Late) Professor Dr. J. N. Kapur
and
Our Families

Preface

Game theory has already proved its tremendous potential for conflict resolution problems in the fields of Decision Theory and Economics. In the recent past, there have been attempts to extend the results of crisp game theory to those conflict resolution problems which are fuzzy in nature e.g. Nishizaki and Sakawa [61] and references cited there in. These developments have lead to the emergence of a new area in the literature called *fuzzy games*. Another area in the fuzzy decision theory, which has been growing very fast is the area of *fuzzy mathematical programming* and its applications to various branches of sciences, Engineering and Management.

In the crisp scenario, there exists a beautiful relationship between two person zero sum matrix game theory and duality in linear programming. It is therefore natural to ask if something similar holds in the fuzzy scenario as well. This discussion essentially constitutes the core of our presentation.

The objective of this book is to present a systematic and focussed study of the application of fuzzy sets to two very basic areas of decision theory, namely *Mathematical Programming* and *Matrix Game Theory*. Apart from presenting most of the basic results available in the literature on these topics, the emphasis here is to understand their natural relationship in a fuzzy environment. The study of duality theory for fuzzy mathematical programming problem plays a key role in understanding this relationship. For this, a theoretical framework of duality in fuzzy mathematical programming and conceptualization of the solution of the fuzzy game is made on the lines of their crisp counterparts. Most of the theoretical results and associated algorithms are illustrated through small numerical examples.

After presenting some basic facts on fuzzy sets and fuzzy arithmetic, the main topics namely fuzzy linear and quadratic programming, fuzzy matrix games, fuzzy bi-matrix games and modality constrained programming are discussed in Chapters 4 to 10. Our presentation is certainly not exhaustive and some topics e.g. fuzzy multi-objective programming and fuzzy multi-objective games have been left deliberately to remain focussed and to keep the book to a reasonable size. Nevertheless these topics are important and therefore appropriate references are provided whenever desirable.

This book is primarily addressed to senior undergraduate students, graduate students and researchers in the area of fuzzy optimization and related topics in the department of Mathematics, Statistics, Operational Research, Industrial Engineering, Electrical Engineering, Computer Science and Management Sciences. Although every care has been taken to make the presentation error free, some errors may still remain and we hold ourselves responsible for that and request that the error if any, be intimated by e-mailing at chandra@maths.iitd.ernet.ac.in (e-mail address of S.Chandra).

In the long process of writing this book we have been encouraged and helped by many individuals. We would first and foremost like to thank Professor Janusz Kacprzyk for accepting our proposal and encouraging us to write this book. We are highly grateful to Professors I. Nishizaki, M. Inuiguchi, J. Ramik, D. Li, T. Maeda and H-C. Wu for sending their reprints / preprints and answering to our queries at the earliest. Their research has certainly been a source of inspiration for us. We would also like to thank the editors and publishers of the journals "Fuzzy Sets and Systems", "Fuzzy Optimization and Decision Making" and "Omega" for publishing our papers in the area of fuzzy linear programming and fuzzy matrix games which constitute the core of this book. We also appreciate our students Ms. Vidyottama Vijay and Ms. Reshma Khemchandani for their tremendous help during the preparation of the manuscript in LATEX and also reading the manuscript from a student point of view. We also acknowledge the book grant provided by IIT Delhi and thank Prof. P.C. Sinha for all help in this regard. Our special thanks are due to Dr. J.L.Gray, Dean, Faculty of Management, University of Manitoba for his encouragement and interest in this work. Last but not the least, we are obliged to Dr. Thomas Ditzinger and Ms Heather King of International Engineering Department, and Mr. Nils Schleusner of Production Department, Springer-Verlag for all their help, cooperation and understanding in the publication of this book.

(Winnipeg), *C.R.Bector*
(New Delhi), *S.Chandra*

Contents

1

Crisp matrix and bi-matrix games: some basic results

1.1 Introduction

There is a vast literature on the theory and applications of (crisp) matrix and bi-matrix games, and some of which have been very well documented in the excellent text books e.g. Jianhua [30], Karlin [31], Parthasarathy and Raghavan [64], and Owen [62]. Therefore in this chapter we only review certain basic results on these topics along with results concerning duality in linear programming. The chapter is divided into six main sections, namely, *duality in linear programming, two-person zero-sum matrix games, linear programming and matrix game equivalence, two person non-zero sum (bi-matrix) games, quadratic programming and bi-matrix games*, and *constrained matrix games.*

1.2 Duality in linear programming

In this section, we will just be quoting certain important results from duality theory of crisp linear programming. We know that, the dual of the standard linear programming problem (called the primal problem)

$$(LP) \qquad \max \quad c^T x$$

subject to,

$$Ax \leq b, x \geq 0,$$

is defined as

$$(LD) \qquad \min \quad b^T y$$

subject to,

$$A^T y \geq c, y \geq 0,$$

where $x \in \mathbb{R}^n$, $y \in \mathbb{R}^m$, $c \in \mathbb{R}^n$, $b \in \mathbb{R}^m$ and A is an $(m \times n)$ real matrix. The above primal-dual pair (LP)-(LD) is symmetric in the sense that the dual of (LD) is (LP). Therefore, out of these two problems (LP) and (LD), anyone could be called primal and the other as its dual. We shall call (LP) as primal and (LD) as its dual.

The following theorems for duality hold between (LP) and (LD).

Theorem 1.2.1 (Weak duality theorem). *Let x be a feasible solution of (LP) and y be a feasible solution of (LD). Then, $c^T x \leq b^T y$.*

Corollary 1.2.1 *Let $\hat{x}$ be a feasible solution of (LP) and $\hat{y}$ be a feasible solution of (LD) such that $c^T \hat{x} = b^T \hat{y}$. Then $\hat{x}$ is an optimal solution of (LP) and $\hat{y}$ is an optimal solution of (LD).*

Theorem 1.2.2 (Duality theorem). *Let $\hat{x}$ be an optimal solution of (LP). Then there exists $\hat{y}$ which is optimal to (LD) and conversely. Further, $c^T \hat{x} = b^T \hat{y}$.*

Theorem 1.2.3 (Existence theorem). *If (LP) is unbounded then (LD) is infeasible, and if (LP) is infeasible and (LD) is feasible, then (LD) is unbounded. Further it is possible that both (LP) and (LD) are infeasible.*

Theorem 1.2.4 (Complementary slackness theorem). *If in any optimal solution of (LP), the slack variable $x^*_{n+i} > 0$, then in every optimal solution of (LD), $y^*_i = 0$. Conversely, if $y^*_i > 0$ in any optimal solution of (LD), then in every optimal solution of (LP) $x^*_{n+i} = 0$, i.e. for a pair of optimal solutions of primal and dual, $x^*_{n+i} y^*_i = 0$ $(i = 1, 2, \ldots, m)$.*

The above theorem can also be stated in the following equivalent way as well.

Let x^* be optimal to (LP) and y^* be optimal to (LD). Then

(i) $\sum_{j=1}^{n} a_{ij} x^*_j < b_i \Rightarrow y^*_i = 0$, and

(ii) $\sum_{i=1}^{m} a_{ij} y^*_i > c_j \Rightarrow x^*_j = 0$.

1.3 Two person zero-sum matrix games

In this section, we present certain basic definitions and preliminaries with regard to two person zero-sum matrix games.

Let $\mathbb{R}^n$ denote the n-dimensional Euclidean space and $\mathbb{R}^n_+$ be its non-negative orthant. Let $A \in R^{m \times n}$ be an $(m \times n)$ real matrix and $e^T = (1, 1, \ldots, 1)$ be a vector of 'ones' whose dimension is specified as per the specific context. By a (crisp) two person zero-sum matrix game G we mean the triplet $G = (S^m, S^n, A)$ where $S^m = \{x \in \mathbb{R}^m_+, e^T x = 1\}$ and $S^n = \{y \in \mathbb{R}^n_+, e^T y = 1\}$. In the terminology of the matrix game theory, S^m(respectively S^n) is called the *strategy space* for *Player I* (respectively *Player II*) and A is called the *pay-off matrix*. Then, the elements of S^m (respectively S^n) which are of the form $x = (0, 0, \ldots, 1, \ldots, 0)^T = e_i$, where 1 is at the ith place (respectively $y = (0, 0, \ldots, 1, \ldots, 0)^T = e_j$, where 1 is at the jth place) are called *pure strategies* for Player I (respectively Player II). If Player I chooses i^{th} pure strategy and Player II chooses j^{th} pure strategy then a_{ij} is the amount paid by Player II to Player I. If the game is zero-sum then $-a_{ij}$ is the amount paid by Player I to Player II i.e. the gain of one player is the loss of other player. The quantity $E(x, y) = x^T A y$ is called the *expected pay-off* of Player I by Player II, as elements of S^m (respectively S^n) can be thought of as a set of all probability distribution over $I = \{1, 2, \ldots, m\}$ (respectively $J = \{1, 2, \ldots, n\}$). It is customary to assume that Player I is a maximizing player and Player II is a minimizing player. The triplet $PG = (I,\ J,\ A)$ is called the *pure form* of the game G , whenever G is being referred as the *mixed extension* of the pure game G. We shall refer to a two person zero-sum game always as G = (S^m, S^n, A) and if the game is in the pure form it will be clear from the context itself. Thus, for us S^m refers to the (mixed) strategy space of Player I, S^n refers to the (mixed) strategy space of Player II, and A refers to the pay-off matrix which introduces the function $E : S^m \times S^n \to \mathbb{R}$ given by $E(x, y) = x^T A y$, called the *expected pay-off function* .

The meaning of the solution of the game $G = (S^m,\ S^n,\ A)$ is best understood in terms of maxmin and minmax principles for Player I and Player II respectively. According to this principle, each player adopts that strategy which results in the best of the worst outcomes. In other words, Player I (the maximizing player) decides to play that strategy which corresponds to the maximum of the minimum gain for his different courses of action. This is known as the *maxmin principle* .

Similarly, Player II (the minimizing player) also likes to play safe and in that case he selects that strategy which corresponds to the minimum of the maximum losses for his different courses of action. This is known as the *minmax principle.*

Employing the maxmin principle for Player I, we obtain $\underline{v} = \max_{x \in S^m} \min_{y \in S^n} (x^T A y)$, called the *lower value* of the game. Similarly the min-max principle for Player II gives $\bar{v} = \min_{y \in S^n} \max_{x \in S^m} (x^T A y)$, called the *upper value* of the game. It is well known that $\bar{v} \geq \underline{v}$. The main result of two-person zero-sum matrix game theory asserts that, in fact, these are equal, i.e $\bar{v} = \underline{v} = v^*$, which is then called the *value of the game.* The following theorem is very useful in this regard.

Theorem 1.3.1 *If there exists* $(x^*, y^*, v^*) \in S^m \times S^n \times \mathbb{R}$ *such that*

(i) $E(x^*, y) \geq v^*, \ \forall\, y \in S^n$, *and,*
(ii) $E(x, y^*) \leq v^*, \ \forall\, x \in S^m$,

then $\bar{v} = v^* = \underline{v}$ *and conversely.*

Definition 1.3.1 (Saddle point). *Let* $E : S^m \times S^n \longrightarrow \mathbb{R}$ *be given by* $E(x, y) = x^T A y$. *The function E is said to have a saddle point* (x^*, y^*) *if* $E(x^*, y) \geq E(x^*, y^*) \geq E(x, y^*)$, $\forall\, x \in S^m$ *and* $\forall\, y \in S^n$.

In view of the above definition we have the following corollary for Theorem 1.3.1.

Corollary 1.3.1 *A necessary and sufficient condition that* $\bar{v} = \underline{v}$ *i.e.* $\min_{y \in S^n} \max_{x \in S^m} x^T A y = \max_{x \in S^m} \min_{y \in S^n} x^T A y$, *is that the function* $E(x, y)$ *has a saddle point* (x^*, y^*). *Here* $v^* = E(x^*, y^*) = \underline{v} = \bar{v}$.

Theorem 1.3.1 leads to the following definition of the solution of the game G.

Definition 1.3.2 (Solution of a game). *Let* $G = (S^m, S^n, A)$ *be the given game. A triplet* $(x^*, y^*, v^*) \in S^m \times S^n \times \mathbb{R}$ *is called a solution of the game G if*

$$E(x^*, y) \geq v^*, \quad \forall \ y \in S^m,$$

and

$$E(x, y^*) \leq v^*, \quad \forall \ x \in S^n.$$

Here x^* *is called an optimal strategy for Player I,* y^* *is called an optimal strategy for Player II, and* v^* *is called the value of the game G .*

Remark 1.3.1. In view of Theorem 1.3.1 and its Corollary 1.3.1, (x^*, y^*, v^*) is a solution of the game G if and only if (x^*, y^*) is a saddle point of E and in that case $v^* = E(x^*, y^*)$. Such a saddle point is guaranteed to exist if $\underline{v} = \bar{v}$ and conversely. Here it may be noted that only the existence of $(\bar{x}, \bar{y}) \in S^m \times S^n$ such that $\min_{y \in S^n} \max_{x \in S^m} x^T A y = \max_{x \in S^m} \min_{y \in S^n} x^T A y = \bar{x}^T A \bar{y}$, is not a sufficient condition in order that $(\bar{x}, \bar{y})$ be a solution of the matrix game G, i.e. this may not imply that $(\bar{x}, \bar{y})$ constitutes an optimal pair of strategies. For example, if $G = (S^2, S^2, A)$ with $A = \begin{bmatrix} 2 & 0 \\ 0 & 2 \end{bmatrix}$ then $\underline{v} = \bar{v} = 1$. Also $x^* = \left(\frac{1}{2}, \frac{1}{2}\right)^T = y^*$ constitutes a saddle point of E and therefore a pair of optimal strategies. However $\bar{x} = \left(\frac{1}{2}, \frac{1}{2}\right)^T$, $\bar{y} = (1,0)^T$ also gives $E(\bar{x}, \bar{y}) = 1$, but $\bar{y}$ is obviously not optimal to Player II. The main reason being that (x^*, y^*) is a saddle point of $E(x, y)$ but $(\bar{x}, \bar{y})$ is not.

Next we answer the basic question regarding the existence of a solution for the game G. The following theorem is very fundamental in this context as it asserts that every two-person zero-sum matrix game G always has a solution.

Theorem 1.3.2 (Fundamental theorem of matrix games). *Let* $G = (S^m, S^n, A)$. *Then* $\min_{y \in S^n} \max_{x \in S^m} x^T A y$ *and* $\max_{x \in S^m} \min_{y \in S^n} x^T A y$ *both exists and are equal.*

Here the problem $\max_{x \in S^m} \min_{y \in S^n} x^T A y$ (respectively $\min_{y \in S^n} \max_{x \in S^m} x^T A y$) is called Player I's (respectively Player II's) problem. If there exists $(i_o, j_o) \in I \times J$ such that $a_{i_o, j} \geq a_{i_o, j_o} \geq a_{i, j_o}$ for all i and j then (i_o, j_o) is called a *pure saddle point* and in that case we say that the game G has a solution in the pure form. In this situation

$$\left(\min_{j \in J} \max_{i \in I} a_{ij}\right) = \left(\max_{i \in I} \min_{j \in J} a_{ij}\right) = a_{i_o, j_o}$$

and therefore i_o gives an optimal pure strategy for Player I, j_o gives an optimal pure strategy for Player II, and a_{i_o, j_o} becomes the value of the game G. In this case it may be noted that a_{i_o, j_o} is the smallest element in the i_o^{th} row and the largest element in the j_o^{th} column.

Thus Theorem 1.3.2 above guarantees that every two person zero-sum matrix game G has a solution. If there is no solution in the pure form then there is certainly a solution in the mixed form. Therefore, the question "How to obtain the solution for this matrix game G?" is to be addressed in the next section.

1.4 Linear programming and matrix game equivalence

We shall now establish an equivalence between two person zero-sum matrix game $G = (S^m,\ S^n,\ A)$ and a pair of primal-dual linear programming problems. This equivalence besides being interesting mathematically, is also very useful as it provides a very efficient way to solve the given game G.

Let us consider the Player I's (respectively Player II's) problem: $\max_{x \in S^m} \min_{y \in S^n}\ x^T A y$ (respectively $\min_{y \in S^n} \max_{x \in S^m}\ x^T A y$). Since S^m and S^n are compact convex sets and for a given x (respectively given y), the function $E(x, y)$ is a linear function of y (respectively x), the $\min_{y \in S^n}\ x^T A y$ $\left(\text{respectively} \max_{x \in S^m} (x^T A y)\right)$ will be attained at an extreme point of S^n (respectively S^m). Therefore for a given $x \in S^m$,

$$\min_{y \in S^n} (x^T A y) = \min_{1 \leq j \leq n} (x^T A e_j),$$

where $e_j = (0, 0, \dots, 1, \dots, 0)^T$ with '1' at the jth place, is the jth pure strategy of Player II. Thus

$$\max_{x \in S^m} \min_{y \in S^n} (x^T A y) = \max_{x \in S^m} \min_{1 \leq j \leq n} \left(\sum_{i=1}^{m} a_{ij} x_i \right).$$

If we now take $v = \min_{1 \leq j \leq n} \left(\sum_{i=1}^{m} a_{ij} x_j \right)$, then the maxmin value for Player I is obtained by solving the following linear programming problem

$$(LP1) \qquad \max \quad v$$
$$\text{subject to,}$$
$$\sum_{i=1}^{m} a_{ij} x_i \geq v, \ (j = 1, 2, \dots, n),$$
$$e^T x = 1,$$
$$x \geq 0.$$

Similarly the minmax value for Player II is obtained as a solution of the following linear programming problem

$$(LD1) \quad \min \quad w$$
$$\text{subject to,}$$
$$\sum_{j=1}^{n} a_{ij} y_j \leq w, \ (i = 1, 2, \ldots, m),$$
$$e^T y = 1,$$
$$y \geq 0,$$

where $w = \max_{1 \leq i \leq m} \left(\sum_{j=1}^{n} a_{ij} y_j \right)$.

Now it can be verified that $(LP1)$ and $(LD1)$ constitute a primal-dual pair of linear programming problems. Since both maxmin and minmax are attained, these two LPPs have optimal solutions ($\bar{x}$ and $\bar{y}$) and therefore by the linear programming duality, the optimal values of $(LP1)$ and $(LD1)$ will be equal. Let this common value be $\bar{v}$. Then the way $(LP1)$ and $(LD1)$ have been constructed, it is obvious that $\sum_{i=1}^{m} a_{ij} \bar{x}_i \geq \bar{v}$, $(j = 1, 2, \ldots, n)$ and $\sum_{j=1}^{n} a_{ij} \bar{y}_j \leq \bar{v}$, $(i = 1, 2, \ldots, m)$ implying that $(\bar{x})^T A y \geq \bar{v}$ for all $y \in S^n$ and $x^T A \bar{y} \leq \bar{v}$ for all $x \in S^m$.

The above discussion then leads to the following equivalence theorem.

Theorem 1.4.1 *The triplet $(\bar{x}, \bar{y}, \bar{v}) \in S^m \times S^n \times \mathbb{R}$ is a solution of the game G if and only if $\bar{x}$ is optimal to $(LP1)$, $\bar{y}$ is optimal to $(LD1)$ and $\bar{v}$ is the common value of $(LP1)$ and its dual $(LD1)$.*

Thus, we have concluded that the matrix game $G = (S^m, S^n, A)$ is equivalent to the primal-dual linear programming problems $(LP1)$-$(LD1)$.

The pair $(LP1)$-$(LD1)$ can further be expressed in the form $(LP2)$-$(LD2)$ where duality is much more obvious and it does not need any checking. For this we need to assume that v^*, the value of the game G, is positive. This assumption can be taken without any loss of generality since matrix games $G = (S^m, S^n, A)$ and $G_1 = (S^m, S^n, A_1)$, $A_1 = (a_{ij} + \alpha)$, $\alpha \in \mathbb{R}$ will have same optimal strategies but different values as v^* and ${v_1}^*$ where ${v_1}^* = v^* + \alpha$. The consequence of the assumption that $v^* > 0$ is that in $(LP1)$ and $(LD1)$ we have $v > 0$ and $w > 0$. Now by defining $x_i' = \frac{x_i}{v}$, $y_j' = \frac{y_j}{w}$ $(i = 1,2,\dots,m,\ j = 1,2,\dots,n)$ and noting that $e^T x' = \frac{1}{v} e^T x = \frac{1}{v}$ and $e^T y' = \frac{1}{w} e^T y = \frac{1}{w}$, and max v (respectively min w) $= \min\left(\frac{1}{v}\right)\left(\text{respectively } \max\left(\frac{1}{w}\right)\right)$, the problems ($LP1$) and ($LD1$) become

$$(LP2) \qquad \begin{array}{ll} \min & e^T x' \\ \text{subject to} & \\ & \sum_{i=1}^{m} a_{ij} x_i' \geq 1,\ (i = 1,2,\dots,m), \\ & x' \geq 0, \end{array}$$

and

$$(LD2) \qquad \begin{array}{ll} \max & e^T y' \\ \text{subject to} & \\ & \sum_{j=1}^{n} a_{ij} y_j' \leq 1,\ (j = 1,2,\dots,n), \\ & y' \geq 0. \end{array}$$

Since ($LP2$)-($LD2$) constitutes a primal-dual pair, it is enough to solve only one of these as the solution of the other will be obtained directly because of the duality theory. Once optimal solution $x^{*'}$ of ($LP2$) and $y^{*'}$ of ($LD2$) are obtained, the value of the game G is obtained as $v^* = w^* = \frac{1}{e^T x^{*'}} = \frac{1}{e^T y^{*'}}$. Also, optimal strategies for Player I and Player II are obtained as $x^* = v^* x^{*'}$ *and* $y^* = v^* y^{*'}$ respectively.

In the above discussion we have constructed a primal-dual pair ($LP1$)-($LD1$) (or ($LP2$)-($LD2$)) for a given general two person zero-sum matrix game G. It is now natural to ask what happens if we are given any general pair of primal-dual linear programming problems say (LP) and (LD). Can we construct an equivalent matrix game G? The answer is in affirmative and that is what we discuss now.

Consider the linear programming problems (LP) together with its dual (LD) as follows

$$(LP) \qquad \begin{array}{ll} \max & c^T x \\ \text{subject to,} & \end{array}$$

$$Ax \leq b,$$
$$x \geq 0,$$

and

(LD) $\quad \min \quad b^T y$

subject to,

$$A^T y \geq c,$$
$$y \geq 0,$$

where $c \in \mathbb{R}^n$, $x \in \mathbb{R}^n$, $b \in \mathbb{R}^m$, $y \in \mathbb{R}^m$, $A = (a_{ij})$ is an $(m \times n)$ real matrix.

Now, consider the matrix game associated with the following $(n + m + 1) \times (n + m + 1)$ skew-symmetric matrix

$$B = \begin{pmatrix} 0 & -A^T & c \\ A & 0 & -b \\ -c^T & b^T & 0 \end{pmatrix}.$$

Since B is a skew-symmetric matrix, the value of the matrix game associated with B is zero and both players have the same optimal strategies. In the following, the matrix game B will mean the matrix game associated with B and indices i and j will run from 1 to m and 1 to n respectively. Also a strategy for either player will be denoted by (x, y, z) where $x \in \mathbb{R}^n$, $y \in \mathbb{R}^m$ and $z \in \mathbb{R}$.

The following result shows that the primal-dual pair (*LP*)-(*LD*) is equivalent to the matrix game B.

Theorem 1.4.2 *Let $\bar{x}$ and $\bar{y}$ be optimal to* (LP) *and* (LD) *respectively. Let* $z^* = \dfrac{1}{1 + \sum\limits_j \bar{x}_j + \sum\limits_i \bar{y}_i}$, $x^* = z^*\bar{x}$, $y^* = z^*\bar{y}$. *Then* (x^*, y^*, z^*) *solves the matrix game* B.

Proof. First we show that $Z^* = (x^*, y^*, z^*)$ will be an optimal strategy for both the players. For this we note that

$$x^* + y^* + z^* = \sum \bar{x}_j z^* + \sum \bar{y}_i z^* + z^* = (1 + \sum \bar{x}_j + \sum \bar{y}_i) z^* = 1,$$

and therefore $(x^*, y^*, z^*) \in S^{m+n+1}$. Now to prove that $Z^* = (x^*, y^*, z^*)$ is an optimal strategy for Player II, we have to show that $BZ^* \leq 0$. But $\bar{x}$ and $\bar{y}$ are solutions of (*LP*) and (*LD*) and therefore by the duality theory

$$A\bar{x} - b \leq 0,$$
$$c - A^T\bar{y} \leq 0,$$
$$-c^T\bar{x} + b^T\bar{y} \leq 0.$$

On multiplying these inequalities by z^*, we have

$$cz^* - A^T y^* \leq 0,$$
$$Ax^* - bz^* \leq 0,$$
$$-c^T x^* + b^T y^* \leq 0,$$

which on writing in matrix form gives $BZ^* \leq 0$.
Now we note that B is skew symmetric and therefore $BZ^* \leq 0$ gives $(Z^*)^T B \geq 0$, which implies that Z^* is an optimal strategy for Player I as well.

Theorem 1.4.3 *Let* (x^*, y^*, z^*) *be an optimal strategy of the matrix game* B *with* $z^* > 0$. *Let* $\bar{x}_j = \dfrac{x^*_j}{z^*}$, $\bar{y}_i = \dfrac{y^*_i}{z^*}$. *Then* $\bar{x}$ *and* $\bar{y}$ *are optimal solutions to* (LP) *and* (LD) *respectively.*

Proof. Since both players have the same optimal strategies, it is sufficient to take $Z^* = (x^*, y^*, z^*)$ as an optimal strategy for either player, say Player II. Similar arguments are valid if Z^* is taken as an optimal strategy for Player I. Therefore, let $Z^* = (x^*, y^*, z^*)$ be an optimal strategy for Player II with $z^* > 0$. Then we have

$$-A^T y^* + cz^* \leq 0,$$
$$Ax^* - bz^* \leq 0,$$
$$-c^T x^* + b^T y^* \leq 0.$$

Now $-A^T y^* + cz^* \leq 0$ gives $A^T\left(\dfrac{y^*}{z^*}\right) \geq c$. Similarly the other two inequalities gives $A\left(\dfrac{x^*}{z^*}\right) \leq b$ and $c^T\left(\dfrac{x^*}{z^*}\right) \geq b^T\left(\dfrac{y^*}{z^*}\right)$. Therefore we have

$$A^T\bar{y} \geq c,$$

$$A\bar{x} \leq b,$$

and

$$c^T\bar{x} \geq b^T\bar{y}.$$

But the first two inequalities imply

$$c^T\bar{x} \le \bar{y}^T A\bar{x} \le \bar{y}^T b = b^T\bar{y},$$

and therefore we have $c^T\bar{x} = b^T\bar{y}$.
This proves that $\bar{x}$ and $\bar{y}$ are optimal for the primal and dual problems respectively.

Thus the equivalence between two person zero-sum matrix game theory and duality in linear programming is complete in the sense that given any general two person zero sum matrix game G, there is a related pair of primal-dual linear programming problems, and given any general pair of primal-dual linear programming problems, there is an associated matrix game B.

1.5 Two person non-zero sum (bi-matrix) games

In the earlier sections, we have studied two person zero-sum games in which the gain of one player is the loss of the other player. But there may be situations in which the interests of two players may not be exactly opposite. Such situations give rise to two person non-zero sum games, also called bi-matrix games . Some well known examples of bi-matrix games are "The Prisoner's Dilemma", "The Battle of Sexes" and "The Bargaining Problem".

A bi-matrix game can be expressed as $BG = (A,\ B,\ S^m,\ S^n)$, where S^m , S^n are as introduced in Section 1.3 and, A and B are $(m \times n)$ real matrices representing the pay-offs to Player I and Player II respectively.

Definition 1.5.1 (Equilibrium solution). *A pair $(x^*,\ y^*) \in S^m \times S^n$ is said to be an equilibrium solution of the bi-matrix game BG if*

$$x^T A y^* \le x^{*T} A y^*,$$

and

$$x^{*T} B y \le x^{*T} B y^*,$$

for all $x \in S^m$ and $y \in S^n$.

Remark 1.5.1. A two person zero sum matrix game $G = (S^m,\ S^n,\ A)$ is a special case of the bi-matrix game BG with $B = -A$. Therefore for $B = -A$, the definition of an equilibrium solution reduces to a saddle point for the two person zero sum game G. This can easily be verified by putting $B = -A$ in Definition 1.5.1.

In the context of bi-matrix game, the following theorem due to Nash [60] is very basic as it guarantees the existence of an equilibrium solution of the bi-matrix game BG.

Theorem 1.5.1 (Nash existence theorem [60]). *Every bi-matrix game $BG = (S^m, S^n, A, B)$ has at least one equilibrium solution.*

Proof. For any $(x, y) \in S^m \times S^n$, let us define

$$c_i(x,y) = \max\,(A_i y - x^T A y,\ 0)$$

and

$$d_j(x,y) = \max\,(x^T B_j - x^T B y,\ 0),$$

where A_i and B_j respectively are the ith row of the matrix A and the jth column of the matrix B.

Then we consider the function $T : S^m \times S^n \longrightarrow S^m \times S^n$ given by $T(x,y) = (x', y')$, where

$$x'_i = \frac{x_i + c_i(x,y)}{1 + \sum_{i=1}^{m} c_i(x,y)}, \quad y'_j = \frac{y_j + d_j(x,y)}{1 + \sum_{j=1}^{n} d_j(x,y)}.$$

Now for $i = 1, 2, \ldots, m$, $c_i(x,y) \geq 0$ and therefore $x'_i \geq 0$. Similarly for $j = 1, \ldots, n$, $d_j(x,y) \geq 0$ and therefore $y'_j \geq 0$. Also it can be verified that $e^T x' = \sum_{i=1}^{m} x'_i = 1$ and $e^T y' = \sum_{j=1}^{n} y'_j = 1$. Hence $x' \in S^m$ and $y' \in S^n$. Further S^m and S^n are compact convex sets and therefore so is $S^m \times S^n$. Now noting that $T : S^m \times S^n \longrightarrow S^m \times S^n$ is a continuous, one to one mapping and $S^m \times S^n$ is a compact convex set, the Brouwer's fixed point theorem asserts that T has at least one fixed point, say (x^*, y^*), $T(x^*, y^*) = (x', y') = (x^*, y^*)$. We shall now show that (x^*, y^*) is an equilibrium solution of the bi-matrix game BG. If possible let (x^*, y^*) be not an equilibrium solution of BG. This means that either there exists some $\bar{x} \in S^m$ such that $\bar{x}^T A y^* > x^{*T} A y^*$ or there exists some $\bar{y} \in S^n$ such that $x^{*T} B \bar{y} > x^{*T} B y^*$. We are here assuming that the first case holds. The proof in the second case is similar. The first case namely, $\bar{x}^T A y^* > x^{*T} A y^*$, implies that there exists some i such that $A_{i.} y^* > x^{*T} A y^*$, which means that $c_i > 0$ for some $i = i_0$. But $c_i \geq 0$ for all i and additionally $c_{i_0} > 0$, and therefore $\sum_{i=1}^{m} c_i > 0$.

Now $x^{*T}Ay^* = \sum_{i=1}^{m}\sum_{j=1}^{n} x_i^* a_{ij} y_j^* = \sum_{i=1}^{m} x_i^* \left(\sum_{j=1}^{n} a_{ij} y_j^*\right)$, is the weighted arithematic mean of m scalars $\sum_{j=1}^{n} a_{ij} y_j^*$ $(i = 1,\ldots,m)$, and therefore $x^{*T}Ay^* \geq \min \left(\sum_j a_{1j} y_j^*, \ldots, \sum_j a_{mj} y_j^*\right) = \sum_j a_{pj} y_j^*$ for some $1 \leq p \leq m$. The above inequality implies that $A_{p.}y^* \leq x^{*T}Ay^*$ and $x_p^* > 0$. Here it may be noted that $x_p^* > 0$ otherwise the corresponding term $\sum_j a_{pj} y_j^*$ will not be present in the minimization.

Therefore $c_p(x^*, y^*) = 0$ which gives $x'_p = \dfrac{x_p^*}{1 + \sum c_{i1}(x^*, y^*)} < x_p^*$ and so $x' \neq x^*$. Similarly in the second case we can show that $y' \neq y^*$. Hence $(x', y') \neq (x^*, y^*)$, which is a contradiction to the fact that (x^*, y^*) is a fixed point. Therefore (x^*, y^*) is an equilibrium solution of the bi-matrix game BG.

1.6 Quadratic programming and bi-matrix game

In Section 1.4 we have shown that every two person zero-sum matrix game $G = (S^m, S^n, A)$ can be solved by solving a suitable primal-dual pair of linear programming problems. Mangasarian and Stone [52] established a somewhat similar result to show that a Nash equilibrium solution of a bi-matrix game BG can be obtained by solving an appropriate quadratic programming problem .

The main result of this section is to obtain the quadratic programming problem that has to be solved in order to obtain an equilibrium solution of the given bi-matrix game BG.

Let us now recall Definition 1.5.1 and note that $(x^*, y^*) \in (S^m \times S^n)$ is a Nash equilibrium solution of the bi-matrix game BG if and only if x^* and y^* simultaneously solve the following problems (P_1) and (P_2), where

(P_1)

$$\begin{aligned} \max \quad & x^T A y^* \\ \text{subject to,} \quad & \\ & e^T x = 1, \\ & x \geq 0, \end{aligned}$$

and

(P_2) $\quad\max\quad (x^*)^T By$

subject to,

$$e^T y = 1,$$
$$y \geq 0.$$

Therefore, a Nash equilibrium solution $(x^*,\ y^*)$ is a pair of strategies x^* and y^* such that

$$x^{*T} Ay^* = \max_x \{\, x^T Ay^* : e^T x - 1 = 0,\ x \geq 0 \,\},$$

and

$$x^{*T} By^* = \max_y \{\, x^{*T} By : e^T y - 1 = 0,\ y \geq 0 \,\}.$$

The following lemma provides necessary and sufficient conditions for an equilibrium solution.

Lemma 1.6.1. *A necessary and sufficient condition that* $(x^*,\ y^*)$ *be an optimal solution of* (P_1) *and* (P_2) *is that there exist scalars* α^* *and* β^* *such that* $(x^*,\ y^*,\ \alpha^*,\ \beta^*)$ *satisfy*

$$\begin{aligned} x^{*T} Ay^* - \alpha^* &= 0 \\ x^{*T} By^* - \beta^* &= 0 \\ Ay^* - \alpha^* e &\leq 0 \\ B^T x^* - \beta^* e &\leq 0 \\ e^T x^* - 1 &= 0 \\ e^T y^* - 1 &= 0 \\ x^* &\geq 0 \\ y^* &\geq 0. \end{aligned}$$

Proof. The proof of the necessary part of the above lemma follows directly by employing the Karush-Kuhn-Tucker conditions to problems (P_1) and (P_2).

Let us now prove the sufficient part of the above lemma; i.e assuming that the given conditions are holding, we have to show that $(x^*,\ y^*)$ is an optimal solution of (P_1) and (P_2). Let $(x,\ y) \in S^m \times S^n$ be arbitrary. Since from the given conditions

$$Ay^* \leq \alpha^* e,$$

we have

$$x^T A y^* \leq \alpha^* x^T e = \alpha^*,$$

i.e

$$x^T A y^* \leq x^{*T} A y^*.$$

Similarly $x^{*T} B y \leq x^{*T} B y^*$. Hence (x^*, y^*) is an equilibrium solution and therefore an optimal solution of (P_1) and (P_2).

Theorem 1.6.1 (Equivalence theorem). *Let $BG = (S^m, S^n, A, B)$ be the given bi-matrix game. A necessary and sufficient condition that (x^*, y^*) be an equilibrium solution of BG is that it is a solution of the following quadratic programming problem.*

$$\begin{array}{ll} \max & x^T(A+B)y - \alpha - \beta \\ \text{subject to} & \\ & Ay - \alpha e \leq 0 \\ & B^T x - \beta e \leq 0 \\ & e^T x - 1 = 0 \\ & e^T y - 1 = 0 \\ & x \geq 0, y \geq 0, \\ & \alpha \in \mathbb{R}, \beta \in \mathbb{R}. \end{array}$$

Further, if $(x^*, y^*, \alpha^*, \beta^*)$ is a solution of the above problem then $\alpha^* = x^{*T} A y^*$, $\beta^* = x^{*T} B y^*$ and $x^{*T}(A+B)y^* - \alpha' - \beta' = 0$.

Proof. Let S be the set of all feasible solutions of the above problem. Then because of Lemma 1.6.1 and Theorem 1.5.1, $S \neq \emptyset$. Now for any arbitrary $(x, y, \alpha, \beta) \in S$

$$\begin{aligned} x^T(A+B)y - \alpha - \beta &= x^T A y + x^T B y - \alpha - \beta \\ &= x^T A y - \alpha e^T x + x^T B y - \beta e^T y \\ &= x^T(Ay - \alpha) + y^T(B^T x - \beta) \leq 0 \end{aligned}$$

and therefore $\max\limits_{x,y,\alpha,\beta} \left(x^T(A+B)y - \alpha - \beta\right) \leq 0$. Now suppose that (x^*, y^*) is an equilibrium solution of the bi-matrix game BG. Then (x^*, y^*) is optimal to (P_1) and (P_2) with $\alpha^* = x^{*T} A y^*$ and $\beta^* = x^{*T} B y^*$. Therefore $(x^*, y^*, \alpha^*, \beta^*) \in S$ and $x^{*T}(A+B)y^* - \alpha^* - \beta^* = 0$. But $\max\limits_{x,y,\alpha,\beta} \left(x^T(A+B)y - \alpha - \beta\right) \leq 0$ and hence $x^{*T}(A+B)y^* - \alpha^* - \beta^* = \max\limits_{x,y,\alpha,\beta} \left(x^T(A+B)y - \alpha - \beta\right) = 0$ which proves the result.

Conversely, let $(x^*, y^*, \alpha^*, \beta^*)$ be a solution of the above quadratic programming problem. Since

$$\max_{x,y,\alpha,\beta} x^T(A+B)y - \alpha - \beta \leq 0,$$

we have

$$x^{*T}(A+B)y^* - \alpha^* - \beta^* \leq 0.$$

But from Nash's Existence Theorem (Theorem 1.5.1) and the Lemma 1.6.1, there exists $(\hat{x},\ \hat{y},\ \hat{\alpha},\ \hat{\beta}) \in S$ such that

$$\hat{x}^T(A+B)\hat{y} - \hat{\alpha} - \hat{\beta} = 0.$$

Therefore the maximum value of the given quadratic programming problem is attained. Since $(x^*,\ y^*,\ \alpha^*,\ \beta^*)$ is optimal, we have

$$x^{*T}(A+B)y^* - \alpha^* - \beta^* = 0.$$

Rest of the proof now follows from the above equation and the given QPP as $x^{*T}Ay^* - \alpha^* = 0$ and $x^{*T}By^* - \beta^* = 0$. Hence $(x^*,\ y^*)$ is an optimal solution of (P_1) and (P_2), which gives that $(x^*,\ y^*)$ is an equilibrium solution of the bi-matrix game BG.

Remark 1.6.2. For $B = -A$, the bi-matrix game $BG = (S^m, S^n, A, B)$ reduces to the two person zero-sum matrix game $G = (S^m,\ S^n,\ A)$. For this case, the quadratic programming problem of Theorem 1.6.1 decomposes itself into following pair of linear programming problems:

(LP) max $-\alpha$

subject to,

$$\begin{aligned} Ay &\leq \alpha e, \\ e^T y &= 1, \\ y &\geq 0, \end{aligned}$$

and

(LD) max $-\beta$

subject to,

$$\begin{aligned} -A^T x &\leq \beta e, \\ e^T x &= 1, \\ x &\geq 0. \end{aligned}$$

Now calling $-\beta$ as λ, we have (LP) and (LD) as

(LP) min α

subject to,

$$Ay \leq \alpha e,$$
$$e^T y = 1,$$
$$y \geq 0,$$

and

(LD) max λ

subject to,

$$A^T x \geq \lambda e,$$
$$e^T x = 1,$$
$$x \geq 0,$$

which is the standard primal-dual pair associated with the two person zero sum game $G = (S^m,\ S^n,\ A)$.

1.7 Constrained matrix games

There are certain matrix game theoretic problems in real life where the strategies of the players are constrained to satisfy general linear inequalities rather than being in S^m or S^n only. These decision problems give rise to *constrained matrix games* which have initially been studied by Charnes [13] and then later in some what more generality by Kawaguchi and Maruyama [34].

Let $S_1 = \{\ x \in \mathbb{R}^m,\ Bx \leq c,\ x \geq 0\ \}, S_2 = \{\ y \in \mathbb{R}^n : D^T y \geq d,\ y \geq 0\ \}$ and $k : S_1 \times S_2 \to \mathbb{R}$ given by $k(x, y) = x^T Ay$, where $x \in \mathbb{R}^m$, $y \in \mathbb{R}^n$, $c \in \mathbb{R}^s$, $d \in \mathbb{R}^t$, $A \in \mathbb{R}^{m \times n}$, $B \in \mathbb{R}^{s \times m}$, and $D \in \mathbb{R}^{n \times t}$. Then the *Constrained matrix games* CG is denoted as $CG = (S_1,\ S_2,\ A)$.

Definition 1.7.1 (Solution of the constrained game CG). *An element $(\bar{x},\ \bar{y}) \in S_1 \times S_2$ is called a solution of the constrained game CG if $(\bar{x},\ \bar{y})$ is a saddle point of the function $k(x, y) = x^T Ay$, $x \in S_1$, $y \in S_2$. In that case the scalar $\bar{x}^T A\bar{y}$ is called the value of the constrained game CG.*

The following is the main theorem of the constrained matrix game theory which, as in the case of usual matrix games, asserts that every constrained matrix game CG is equivalent to two linear programming problems (CLP) and (CLD) which are dual to each other, where

(CLP) max $d^T u$

subject to,

$$-A^T x + Du \leq 0,$$
$$Bx \leq c,$$
$$x, u \geq 0,$$

and

(*CLD*) $\quad\min\quad c^T v$

subject to,

$$-Ay + B^T v \geq 0,$$
$$D^T y \geq d,$$
$$y, v \geq 0.$$

Theorem 1.7.1 *An element* $(\bar{x},\ \bar{y}) \in S_1 \times S_2$ *is a solution of the constrained game* $CG = (S_1,\ S_2,\ A)$ *if and only if there exist* $\bar{u} \in \mathbb{R}^t$, $\bar{v} \in \mathbb{R}^s$ *such that* $(\bar{x},\ \bar{u})$ *and* $(\bar{y},\ \bar{v})$ *are optimal to the mutually dual pair of linear programming problems* (*CLP*)-(*CLD*)

Proof. Let us first assume that $(\bar{x},\ \bar{y}) \in S_1 \times S_2$ is a solution of the constrained game CG. This by definition implies that $k(x, \bar{y}) \leq k(\bar{x}, \bar{y}) \leq k(\bar{x}, y)$, for all $x \in S_1$ and $y \in S_2$. But then the left hand side of the above inequality means that $\bar{x}$ is an optimal solution of the linear programming problem

$(LP(\bar{y}))$ $\quad\max\quad x^T(A\bar{y})$

subject to,

$$Bx \leq c,$$
$$x \geq 0.$$

Hence, by the duality theorem, there exists $\bar{v} \in \mathbb{R}^s$ which is optimal to the dual $(LD(\bar{y}))$ where

$(LD(\bar{y}))$ $\quad\min\quad c^T v$

subject to,

$$B^T v \geq A\bar{y},$$
$$v \geq 0.$$

Also $c^T \bar{v} = (\bar{x})^T A\bar{y}$.

Similarly, the right hand side of the saddle point inequality gives that $\bar{y}$ is an optimal solution of the linear programming problem

$(LP(\bar{x}))$ $\quad\min\quad (\bar{x}^T A)y$

subject to,

$$D^T y \geq d,$$
$$y \geq 0,$$

Hence, by the duality theorem, there exists $\bar{u} \in \mathbb{R}^t$ which is optimal to the dual $(LD(\bar{x}))$, where

$$
\begin{aligned}
(LD(\bar{x})) \qquad & \max \quad d^T u \\
& \text{subject to,} \\
& \quad Du \leq A^T \bar{x}, \\
& \quad u \geq 0.
\end{aligned}
$$

Also $d^T \bar{u} = \bar{x}^T A \bar{y}$.
Now looking at the linear programming problem $(LD(\bar{y}))$, and also noting that $\bar{y} \in S_2$ and $\bar{x}^T A \bar{y} = c^T \bar{v}$, we note that $(\bar{y}, \bar{v})$ is optimal to the linear programming problem

$$
\begin{aligned}
& \min \quad c^T v \\
& \text{subject to,} \\
& \quad -Ay + B^T v \geq 0, \\
& \quad D^T y \geq d, \\
& \quad y, v \geq 0,
\end{aligned}
$$

which is same as the problem (CLD).
Similarly from $(LP(\bar{x}))$ and the fact that $x \in S_1$ and $\bar{x}^T A \bar{y} = d^T \bar{u}$, we note that $(\bar{x}, \bar{u})$ is optimal to the linear programming problem

$$
\begin{aligned}
& \max \quad d^T u \\
& \text{subject to,} \\
& \quad -A^T x + Du \leq 0, \\
& \quad Bx \leq c, \\
& \quad x, u \geq 0,
\end{aligned}
$$

which is the same as the problem (CLP).
It is simple to verify that problems (CLP) and (CLD) are dual to each other and $d^T \bar{u} = c^T \bar{v} = \bar{x}^T A \bar{y}$.
Conversely, suppose that corresponding to $(\bar{x}, \bar{y}) \in S_1 \times S_2$, there exist $\bar{u} \in \mathbb{R}^s$ and $\bar{v} \in \mathbb{R}^t$ such that $(\bar{x}, \bar{u})$ and $(\bar{y}, \bar{v})$ are optimal to (CLP) and (CLD) respectively. We shall show that $(\bar{x}, \bar{y})$ is a saddle point of $k(x, y) = x^T Ay$, $x \in S_1$, $y \in S_2$. For this we observe that from the given hypothesis

$$
\begin{aligned}
-A^T \bar{x} + D\bar{u} &\leq 0, \\
B\bar{x} &\leq c, \\
-A\bar{y} + B^T \bar{v} &\geq 0,
\end{aligned}
$$

$$D^T\bar{y} \geq d,$$
$$c^T\bar{u} = d^T\bar{v},$$
$$\bar{x}, \bar{y}, \bar{u}, \bar{v} \geq 0.$$

Therefore,

$$x^T A\bar{y} \leq x^T B^T \bar{v} = (Bx)^T \bar{v} \leq c^T \bar{v},\ x \in S_1,$$

and

$$\bar{x}^T Ay \geq \bar{u}^T D^T y = \bar{u}^T d = d^T \bar{u},\ y \in S_2.$$

The above inequalities imply

$$x^T A\bar{y} \leq c^T \bar{v} = d^T \bar{u} \leq \bar{x}^T Ay,\ x \in S_1,\ y \in S_2.$$

But

$$x^T A\bar{y} \leq c^T \bar{v},\ x \in S_1$$

gives

$$\bar{x}^T A\bar{y} \leq c^T \bar{v}.$$

Similarly

$$\bar{x}^T Ay \geq d^T \bar{u},$$

gives

$$\bar{x}^T A\bar{y} \geq d^T \bar{u},$$

and hence

$$c^T \bar{v} = d^T \bar{u} = \bar{x}^T A\bar{y}.$$

Thus for all $x \in S_1,\ y \in S_2,$ we have $x^T A\bar{y} \leq \bar{x}^T A\bar{y} \leq \bar{x}^T Ay$.

1.8 Conclusions

In this chapter we have presented certain basic results on duality in linear programming, two person zero-sum matrix games, and bi-matrix games. The discussion on Karush Kuhn-Tucker (K.K.T) conditions and duality in quadratic programming has not been included here, and for that we may have to refer to excellent texts like Bazaraa, Sherali and Shetty [2], and Mangasarian [53].

2

Fuzzy sets

2.1 Introduction

The purpose of this chapter is to review the basic definitions and results on fuzzy sets and related topics. The chapter is divided into six main sections, namely, *basic definitions and set theoretic operations, α-cuts and their properties, fuzzy relations, Zadeh's extension principle, convex fuzzy sets, triangular norms (t-norms)* and *triangular conorms (t-conorms)*.

Most of the results in this chapter are without proofs. Some appropriate references for this chapter are Dumitrescu, Lazzerini and Jain [17], Klir and Yuan [35], Lin and Lee [45] and Zimmermann [91].

2.2 Basic definitions and set theoretic operations

In this section we introduce some of the basic terminologies of fuzzy set theory and present various set theoretic operations.

Definition 2.2.1 (Fuzzy set). *Let X be the universe whose generic element be denoted by x. A fuzzy set A in X is a function $A : X \longrightarrow [0,1]$.*

We frequently use μ_A for the function A and say that the fuzzy set A is characterized by its *membership function* $\mu_A : X \longrightarrow [0,1]$ which associates with each x in X, a real number $\mu_A(x)$ in [0,1]. The value $\mu_A(x)$ at x represents the *grade of membership* of x in A and is interpreted as the degree to which x belongs to A. Thus the *closer* the value of $\mu_A(x)$ is to 1, the *more* x belongs to A.

A crisp or ordinary subset A of X can also be viewed as a fuzzy set in X with membership function as its characteristic function, i.e.

$$\mu_A(x) = \begin{cases} 0, x \notin A, \\ 1, x \in A. \end{cases}$$

Sometimes a fuzzy set A in X is denoted by listing the ordered pairs $(x, \mu_A(x))$, where the elements with zero degree are usually not listed. Thus a fuzzy set A in X can also be represented as $A = \{(x, \mu_A(x))\}$ where $x \in X$ and $\mu_A : X \longrightarrow [0,1]$. As $\mu_A : X \longrightarrow [0,1]$, the following definitions are natural in this context.

Definition 2.2.2 (Support of a fuzzy set). *Let A be a fuzzy set in X. Then the support of A, denoted by $S(A)$, is the crisp set given by*

$$S(A) = \{x \in X : \mu_A(x) > 0\}.$$

Definition 2.2.3 (Normal fuzzy set). *Let A be a fuzzy set in X. The height $h(A)$ of A is defined as*

$$h(A) = \sup_{x \in X} \mu_A(x).$$

If $h(A) = 1$, then the fuzzy set A is called a normal fuzzy set, otherwise it is called subnormal. If $0 < h(A) < 1$, then the subnormal fuzzy set A can be normalized, i.e. it can be made normal by redefining the membership function as $\mu_A(x)/h(A)$, $x \in X$.

Example 2.2.1. Let X={30, 50, 70, 90} be possible speeds (kmph) at which cars can cruise over long distances. Then the fuzzy set A of "comfortable speeds for long distances" may be defined subjectively by a certain individual as

$$\begin{aligned} \mu(x = 30) &= 0.5, \\ \mu(x = 50) &= 0.8, \\ \mu(x = 70) &= 1, \\ \mu(x = 90) &= 0.4, \end{aligned}$$

where $\mu(\cdot)$ is the membership function of the fuzzy set A of X. This fuzzy set can also be represented as A={(30, 0.5), (50, 0.8), (70, 1), (90, 0.4)}.

Example 2.2.2. Let X be the set of reals $\mathbb{R}$ and A be the fuzzy set of real numbers which are in the "vicinity" of 15. Then a precise though subjective characterization of A by specifying μ_A as a function on $\mathbb{R}$ can be given as

$$\mu_A(x) = \left(1 + (x - 15)^4\right)^{-1}.$$

Some representative values of this function will be

$$\mu_A(12) = 0.01,$$
$$\mu_A(14) = 0.5,$$
$$\mu_A(14.5) = 0.9,$$
$$\mu_A(15) = 1,$$
$$\mu_A(15.5) = 0.9,$$
$$\mu_A(17) = 0.06.$$

Here the fuzzy set of both Examples 2.2.1 and 2.2.2 are normal. However let in the Example 2.2.1, $\mu(x = 70) = 0.9$ instead of 1. Then the fuzzy set A of "comfortable speeds for long distances" is a subnormal fuzzy set of height 0.9 which can be normalized by dividing each $\mu_A(x)$ by 0.9.

Next we proceed to define certain standard set theoretic operations for fuzzy sets. We shall have more discussions on these operations in Section 2.7 where t-norms and t-conorms are introduced. In the following let A and B be two fuzzy sets in X.

Definition 2.2.4 (Empty fuzzy set). *A fuzzy set A is empty if its membership function is identically zero, i.e. $\mu_A(x) = 0$ for all $x \in X$.*

Definition 2.2.5 (Subset). *A fuzzy set A is a subset of a fuzzy set B or A is contained in B if $\mu_A(x) \leq \mu_B(x)$ for all $x \in X$. This is denoted as $A \subseteq B$.*

Definition 2.2.6 (Equality of fuzzy sets). *Two fuzzy sets A and B are said to be equal if $A \subseteq B$ and $B \subseteq A$, i.e. $\mu_A(x) = \mu_B(x)$ for all $x \in X$.*

Definition 2.2.7 (Standard complement). *The standard complement of a fuzzy set A is another fuzzy set, denoted by A', whose membership function is defined as $\mu_{A'}(x) = 1 - \mu_A(x)$ for all $x \in X$.*

Definition 2.2.8 (Standard union). *The standard union of two fuzzy sets A and B is a fuzzy set C whose membership function is given by*

$$\mu_C(x) = \max\left(\mu_A(x), \mu_B(x)\right)$$

for all $x \in X$. This we express as $C = A \cup B$.

Definition 2.2.9 (Standard intersection). *The standard intersection of two fuzzy sets A and B is a fuzzy set D whose membership function is given by*

$$\mu_D(x) = \min\left(\mu_A(x), \mu_B(x)\right)$$

for all $x \in X$. This we express as $D = A \cap B$.

Due to the associativity of *min* and *max* operations, definitions of union and intersection can be extended to any finite number of fuzzy sets in an obvious manner. Here it can be verified that the following properties of crisp sets hold for fuzzy sets as well:

(i) $A \cup B = B \cup A$ (commutativity)
(ii) $(A \cup B) \cup C = A \cup (B \cup C)$
$(A \cap B) \cap C = A \cap (B \cap C)$ (associativity)
(iii) $(A \cup B)' = A' \cap B'$
$(A \cap B)' = A' \cup B'$ (De Morgan's laws)
(iv) $A \cup (B \cap C) = (A \cup B) \cap (A \cup C)$
$A \cap (B \cup C) = (A \cap B) \cup (A \cap C)$ (distributive laws).

The following two properties of crisp sets *do not* hold for fuzzy sets

(i) $A \cap A' = \emptyset$ (law of contradiction),
(ii) $A \cup A' = X$ (law of excluded middle).

2.3 α-Cuts and their properties

In the following, certain crisp sets, called *α-cuts*, are introduced for a given fuzzy set A in X. These (crisp) sets play an important role in the study of fuzzy set theory because every fuzzy set A in X can uniquely be represented by a family of such sets associated with A. Further, the employment of the notion of α-cuts becomes very handy and convenient in the study of the fuzzy arithmetic, to be studied in Chapter 3.

Definition 2.3.1 (α-cut). *Let A be a fuzzy set in* X *and $\alpha \in (0,1]$. The α-cut of the fuzzy set A is the crisp set A_α given by*

$$A_\alpha = \{x \in X : \mu_A(x) \geq \alpha\}.$$

One can check that in the Example 2.2.1, $A_{0.5} = \{30, 50, 70\}$ and $A_{0.8} = \{50, 70\}$.

From the definition of α-cut, it immediately follows that for any fuzzy set A and pair $\alpha_1, \alpha_2 \in (0,1]$, $\alpha_1 \leq \alpha_2$, one has $A_{\alpha_2} \subseteq A_{\alpha_1}$. Therefore, all α-cuts of any fuzzy set form families of crisp sets which can be used to represent a given fuzzy set A in X. This is summarized in the form of following theorems.

Theorem 2.3.1 *Let A be a fuzzy set in X with the membership function $\mu_A(x)$. Let A_α be the α-cuts of A and $\chi_{A_\alpha}(x)$ be the characteristic function of the crisp set A_α for $\alpha \in (0,1]$. Then*

$$\mu_A(x) = \sup_{\alpha \in (0,1]} \left(\alpha \wedge \chi_{A_\alpha}(x)\right), \ x \in X.$$

Proof. Since $\chi_{A_\alpha}(x)$ is the characteristic function of the crisp set A_α it takes the value 1 if $x \in A_\alpha$ and it takes the value 0 if $x \notin A_\alpha$. Therefore combining them with the definition of α-cut we have

$$x \in A_\alpha \Longrightarrow \chi_{A_\alpha}(x) = 1 \text{ (and also } \mu_A(x) \geq \alpha)$$

and

$$x \notin A_\alpha \Longrightarrow \chi_{A_\alpha}(x) = 0 \text{ (and also } \mu_A(x) < \alpha).$$

Now

$$\begin{aligned}
\sup_{\alpha \in (0,1]} \left(\alpha \wedge \chi_{A_\alpha}(x)\right) &= \left(\sup_{\alpha \in (0,\, \mu_A(x)]} \left(\alpha \wedge \chi_{A_\alpha}(x)\right) \right) \vee \left(\sup_{\alpha \in (\mu_A(x),\, 1]} \left(\alpha \wedge \chi_{A_\alpha}(x)\right) \right) \\
&= \left(\sup_{\alpha \in (0,\, \mu_A(x)]} (\alpha \wedge 1) \right) \vee \left(\sup_{\alpha \in (\mu_A(x),\, 1]} (\alpha \wedge 0) \right) \\
&= \sup_{\alpha \in (0,\, \mu_A(x)]} \alpha \\
&= \mu_A(x).
\end{aligned}$$

Remark 2.3.1. Given a fuzzy set A in X, one can consider a special fuzzy set, denoted by αA_α for $\alpha \in (0,1]$, whose membership function is defined as

$$\mu_{\alpha A_\alpha}(x) = \left(\alpha \wedge \chi_{A_\alpha}(x)\right), \ x \in X.$$

Also, one may introduce the set

$$\Lambda_A = \{\alpha : \ \mu_A(x) = \alpha \text{ for some } x \in X\},$$

called the level set of A. Then the above theorem states that the fuzzy set A can be expressed in the form

$$A = \bigcup_{\alpha \in \Lambda_A} (\alpha A_\alpha),$$

where $\bigcup$ denotes the standard fuzzy union. This result is called the *resolution principle* of fuzzy sets. The essence of *resolution principle* is that a fuzzy set A can be decomposed into fuzzy sets αA_α, $\alpha \in (0,1]$. Looking from a different angle, it tells that a fuzzy set A in X can be retrieved as a union of its αA_α sets, $\alpha \in (0,1]$. This is called the *representation theorem* of fuzzy sets. Thus the *resolution principle* and *representation theorem* are the two sides of the same coin as both of them essentially tell that a fuzzy set A in X can always be expressed in terms of its α-cuts without explicitly resorting to its membership function $\mu_A(x)$.

2.4 Convex fuzzy sets

The notion of convexity of crisp sets in $\mathbb{R}^n$ plays an important role in crisp mathematical programming and game theory. Here this notion of convexity is extended to fuzzy sets in $\mathbb{R}^n$ and some of their properties are discussed. The convexity of fuzzy sets is very crucial to the very definition of a *fuzzy number* and related *fuzzy arithmetic* as will be observed in the next chapter.

Definition 2.4.1 (Convex fuzzy set). *A fuzzy set A in $\mathbb{R}^n$ is said to be a convex fuzzy set if its α-cuts A_α are (crisp) convex sets for all $\alpha \in (0,1]$.*

Definition 2.4.2 (Bounded fuzzy set). *A fuzzy set A in $\mathbb{R}^n$ is said to be a bounded fuzzy set if its α-cuts A_α are (crisp) bounded sets for all $\alpha \in (0,1]$.*

A fuzzy set A in $\mathbb{R}^n$ which is both bounded and convex is called *bounded convex fuzzy set.* The following result gives an equivalent definition of a convex fuzzy set.

Theorem 2.4.1 *A fuzzy set A in $\mathbb{R}^n$ is a convex fuzzy set if and only if for all $x_1,\ x_2 \in \mathbb{R}^n$ and $0 \leq \lambda \leq 1$,*

$$\mu_A\Big(\lambda x_1 + (1-\lambda)x_2\Big) \geq \min\Big(\mu_A(x_1),\ \mu_A(x_2)\Big).$$

Proof. Let A be a convex fuzzy set in the sense of Definition 2.4.1. Let $\alpha = \mu_A(x_1) \leq \mu_A(x_2)$. Then $x_1 \in A_\alpha$, $x_2 \in A_\alpha$ and also $\lambda x_1 + (1-\lambda)x_2 \in A_\alpha$ by the convexity of A_α. Therefore

$$\mu_A(\lambda x_1 + (1-\lambda)x_2) \geq \alpha = \min(\mu_A(x_1),\ \mu_A(x_2)).$$

Conversely, if the membership function μ_A of the fuzzy set A satisfies the inequality of Theorem 2.4.1, then taking $\alpha = \mu_A(x_1)$, A_α may be regarded as set of all points x_2 for which $\mu_A(x_2) \geq \alpha = \mu_A(x_1)$. Therefore for all $x_1,\ x_2 \in A_\alpha$,

$$\mu_A(\lambda x_1 + (1-\lambda)x_2) \geq \min\Big(\mu_A(x_1),\ \mu_A(x_2)\Big) = \mu_A(x_1) = \alpha,$$

which implies that $\lambda x_1 + (1-\lambda)x_2 \in A_\alpha$. Hence A_α is a convex set for every $\alpha \in (0,1]$.

Remark 2.4.1. The convexity of a fuzzy set does not mean that its membership function μ_A is a convex function in the crisp sense. In fact, membership functions of convex fuzzy sets are functions that, according to standard definitions in the mathematical programming literature are quasi-concave (a generalization of the usual concave function) and not convex.

The following diagrams depict a convex fuzzy set and also a nonconvex fuzzy set.

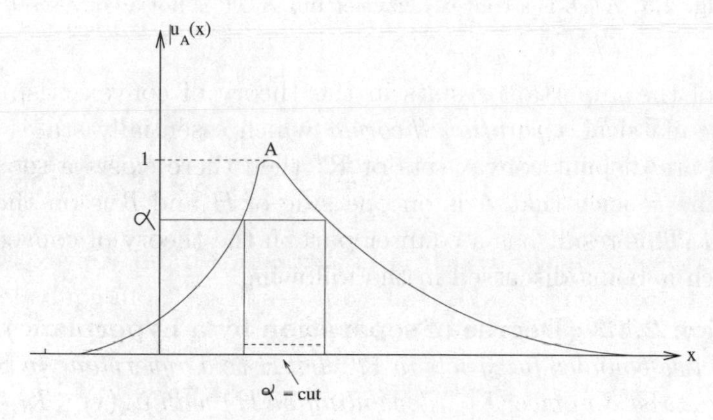

Fig. 2.1. a convex fuzzy set

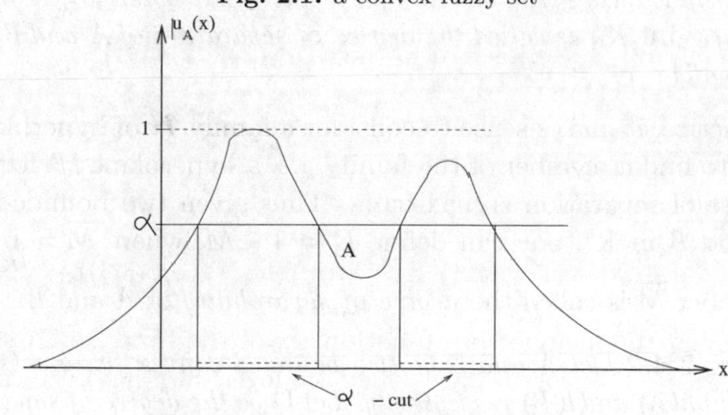

Fig. 2.2. a nonconvex fuzzy set

Remark 2.4.2. It can be easily verified that if A and B are two convex fuzzy sets in $\mathbb{R}^n$ then so is their intersection. However the union of A and B need not be a convex fuzzy set. This is depicted in the below given Figure 2.3.

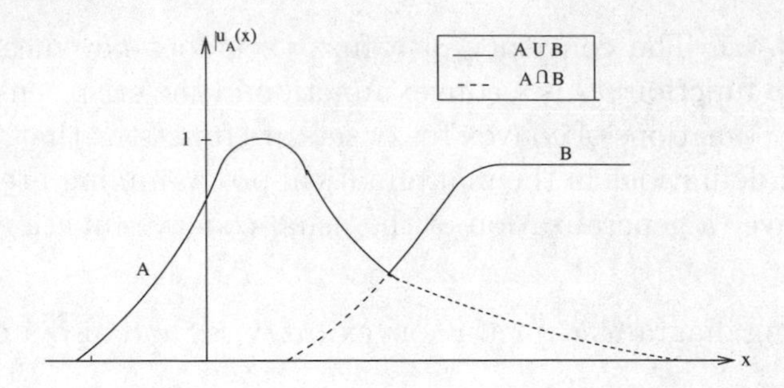

Fig. 2.3. $A \cap B$ is a convex fuzzy set but $A \cup B$ is not a convex set.

One of the important results in the theory of convex crisp sets of $\mathbb{R}^n$ is the classical *separation theorem* which essentially states that if A and B are disjoint convex sets of $\mathbb{R}^n$ then there exists a separating hyperplane H such that A is on one side of H and B is on the other side of H. This result has a counter part in the theory of convex fuzzy sets which is being discussed in the following.

Definition 2.4.3 (Degree of separation by a hyperplane). *Let A and B be two bounded fuzzy sets in $\mathbb{R}^n$. Let H be a hyperplane in $\mathbb{R}^n$ such that there exist a number K_H (depending on H) with $\mu_A(x) \leq K_H$ on one side of H and $\mu_B(x) \leq K_H$ on the other side of H. Then $D_H = 1 - M_H$, where $M_H = \inf K_H$ is called the degree of separation of A and B by the hyperplane H.*

In practice, it makes sense to consider a family $\mathcal{H}$ of hyperplanes H and aim to find a member of the family, i.e. a hyperplane H^* for which the degree of separation is maximum. Thus given two bounded fuzzy sets A and B in $\mathbb{R}^n$, one can define $D = 1 - \overline{M}$, where $\overline{M} = \inf\limits_{H \in \mathcal{H}} M_H$. This number $\overline{M}$ is called the *degree of separability* of A and B.

Theorem 2.4.2 *Let A and B be two bounded convex fuzzy sets in $\mathbb{R}^n$ with height $h(A)$ and $h(B)$ respectively. Let D be the degree of separability of A and B. Then $D = 1 - h(A \cap B)$.*

This theorem of Zadeh [89], which is not being proved here, essentially tells that the highest degree of separation (i.e. the degree of separability) equals $1 - h(A \cap B)$ and this can be achieved by a hyperplane $H^* \in \mathcal{H}$.

2.5 Zadeh's extension principle

The *extension principle of Zadeh* is a very important tool in the fuzzy set theory which provides a procedure to *fuzzify* a crisp function or possibly a crisp relation. This type of fuzzification helps to study mathematical relationships between fuzzy entities and thereby facilitates the study of various real life fuzzy systems. One direct application of this principle will be seen in the next chapter on *fuzzy numbers and fuzzy arithmetic*.

Let $f : X \to Y$ be a crisp function and $F(X)$ (respectively $F(Y)$) be the set of all fuzzy sets (called *fuzzy power set*) of X (respectively Y). The function $f : X \to Y$ induces two functions $f : F(X) \to F(Y)$ and $f^{-1} : F(Y) \to F(X)$, and the extension principle of Zadeh gives formulas to compute the membership function of fuzzy sets $f(A)$ in Y (respectively $f^{-1}(B)$ in X) in terms of membership function of fuzzy set A in X (respectively B in Y).

Definition 2.5.1 (Zadeh's extension principle). *In terms of the notation introduced above, the extension principle of Zadeh states that*

(i) $\mu_{f(A)}(y) = \sup_{x\in X,\ f(x)=y} \big(\mu_A(x)\big)$, *for all* $A \in F(X)$, *and*

(ii) $\mu_{f^{-1}(B)}(x) = \mu_B\big(f(x)\big)$, *for all* $B \in F(Y)$.

Sometimes the function f maps n-tuple in X to a point in Y i.e. $X = X_1 \times X_2 \times \ldots \times X_n$ and $f : X \to Y$ given by $y = f(x_1, x_2, \ldots, x_n)$. Let $A_1, A_2, \ldots, A_n$ be n fuzzy sets in $X_1, X_2, \ldots, X_n$ respectively. The extension principle of Zadeh allows to extend the crisp function $y = f(x_1, x_2, \ldots, x_n)$ to act on n fuzzy subsets of X, namely $A_1, A_2, \ldots, A_n$ such that $B = f(A_1, A_2, \ldots, A_n)$.

Here the fuzzy set B is defined by

$$B = \Big\{ \big(y, \mu_B(y)\big) : y = f(x_1, \ldots, x_n),\ (x_1, \ldots, x_n) \in X_1 \times \ldots \times X_n \Big\}$$

and

$$\mu_B(y) = \sup_{x\in X,\ y=f(x)} \min \big(\mu_{A_1}(x_1), \ldots, \mu_{A_n}(x_n)\big).$$

Example 2.5.1. (Lin and Lee [45]). Let $X = \{-2,\ -1,\ 0,\ 1,\ 2\}$ and A be a fuzzy set in X given by $A = \{(-1, 0.5),\ (0, 0.8),\ (1, 1),\ (2, 0.4)\}$. Let the function $f : X \to \mathbb{R}$ be given by $y = f(x) = x^2$. From this data, one can make the following calculations.

x	$\mu_A(x)$	$y = x^2$	$\mu_B(y) = \mu_{f(A)}(y)$
−1	0.5	1	$\max\{0.5, 1.0\} = 1.0$
0	0.8	0	0.8
1	1	1	$\max\{0.5, 1.0\} = 1.0$
2	0.4	4	0.4

Here $x = 1$ and $x = -1$ both are mapped to the point $y = 1$ under the mapping $y = x^2$ and therefore the membership grade of $y = 1$ is taken as $\max(0.5,\ 1.0) = 1.0$. Therefore

$$B = f(A) = \Big\{(1,1),\ (0,0.8),\ (4,0.4)\Big\}.$$

2.6 Fuzzy relations

Let X and Y be two crisp sets and $X \times Y$ be their cross product. A crisp binary relation $\mathcal{R}$ is a subset of $X \times Y$. It indicates the presence or absence of certain association between the elements of sets X and Y. If one allows the presence of this association to be of varying degree between 0 and 1 then a binary fuzzy relation is obtained.

Definition 2.6.1 (Binary fuzzy relation). *A binary fuzzy relation $\mathcal{R}(X, Y)$ on $X \times Y$ is defined as*

$$\mathcal{R}(X, Y) = \Big\{\big((x, y),\ \mu_{\mathcal{R}}(x, y)\big) : (x, y) \in X \times Y\Big\},$$

where $\mu_{\mathcal{R}} : X \times Y \to [0, 1]$ is a grade of membership function. If $X = Y$ then $\mathcal{R}(X, Y)$ is called a binary fuzzy relation on X.

Although an *n-ary fuzzy relation* on a product space $X = X_1 \times X_2 \times \ldots \times X_n$ may be defined by an n-variate membership function $\mu_{\mathcal{R}}(x_1, \ldots, x_n)$ in the similar way, but the discussion here will be confined to binary fuzzy relations only.

Given a binary fuzzy relation $\mathcal{R}(X, Y)$, the fuzzy set *dom* $\mathcal{R}$ in X whose membership function is defined by

$$\text{dom } \mathcal{R}(x) = \max_{y \in Y} \mathcal{R}(x, y),$$

is called the *domain* of $\mathcal{R}$. Similarly the fuzzy set *ran* $\mathcal{R}$ in Y whose membership function is defined by

$$\text{ran } \mathcal{R}(y) = \max_{x \in X} \mathcal{R}(x, y),$$

is called the *range* of $\mathcal{R}$. Further, the number $h(\mathcal{R})$ is given by

$$h(\mathcal{R}) = \max_{y \in Y} \max_{x \in X} \mathcal{R}(x, y),$$

is called the *height* of the fuzzy relation $\mathcal{R}$. Here we are using the same notation for the fuzzy relation $\mathcal{R}$ and its membership function $\mu_{\mathcal{R}}$.

If $X = \{x_1, x_2, \ldots, x_n\}$ and $Y = \{y_1, y_2, \ldots, y_n\}$, then the binary fuzzy relation $\mathcal{R}(X, Y)$ can be represented by an $(m \times n)$ matrix $\mathcal{R}(X, Y) = \big(\mu_{\mathcal{R}}(x_i, y_j)\big)_{m \times n}$, called the *fuzzy matrix* of $\mathcal{R}$.

Similar to the inverse of a crisp relation, the *inverse* of a binary fuzzy relation $\mathcal{R}(X, Y)$ on $X \times Y$, denoted by $\mathcal{R}^{-1}(Y, X)$, is a relation on $Y \times X$ given by $\mathcal{R}^{-1}(y, x) = \mathcal{R}(x, y)$ for all $x \in X$ and all $y \in Y$. Here $(\mathcal{R}^{-1})^{-1} = \mathcal{R}$ and, in case the relation $\mathcal{R}(X, Y)$ is represented by a fuzzy matrix M, the matrix for $\mathcal{R}^{-1}$ will be M^T.

Example 2.6.1. Consider the binary fuzzy relation $\mathcal{R}$ on reals $\mathbb{R}$ described by "y is much larger than x". Then a subjectively chosen membership function for $\mathcal{R}$ could be taken as

$$\mu_{\mathcal{R}}(x, y) = \begin{cases} 0 & , y \le x, \\ \left((1 + (y - x)^{-2}\right)^{-1} & , y > x. \end{cases}$$

Definition 2.6.2 (Standard fuzzy composition). *Let $\mathcal{P}(X, Y)$ be a binary fuzzy relation on $X \times Y$ and $\mathcal{Q}(Y, Z)$ be a binary fuzzy relation on $Y \times Z$. Then the standard fuzzy composition of $\mathcal{P}(X, Y)$ and $\mathcal{Q}(Y, Z)$, denoted by $\mathcal{P}(X, Y) \circ \mathcal{Q}(Y, Z)$, is a binary fuzzy relation $\mathcal{R}(X, Z)$ on $X \times Z$ defined by*

$$\mathcal{R}(x, z) = (\mathcal{P} \circ \mathcal{Q})(x, z) = \max_{y \in Y} \Big(\min \big(\mathcal{P}(x, y), \mathcal{Q}(y, z)\big)\Big)$$

for $x \in X$ and $z \in Z$. This standard composition is also called the max-min composition of relation $\mathcal{P}$ and $\mathcal{Q}$.

Using the above definition it can be verified that similar to crisp binary relations

$$\big(\mathcal{P}(X, Y) \circ \mathcal{Q}(Y, Z)\big)^{-1} = \mathcal{Q}^{-1}(Z, Y) \circ \mathcal{P}^{-1}(Y, X),$$

and

$$\big(\mathcal{P}(X, Y) \circ \mathcal{Q}(Y, Z)\big) \circ \mathcal{R}(Z, W) = \mathcal{P}(X, Y) \circ \big(\mathcal{Q}(Y, Z) \circ \mathcal{R}(Z, W)\big).$$

In this context it may be noted that the standard composition is, in general, not commutative, i.e.

$$\mathcal{P}(X,Y) \circ \mathcal{Q}(Y,Z) \neq \mathcal{Q}(Y,Z) \circ \mathcal{P}(X,Y).$$

Here, of course it is assumed that $Z = X$, otherwise $\mathcal{Q} \circ \mathcal{P}$ is not well defined. If X and Y are discrete sets then the composition of binary fuzzy relation can be performed very conveniently by using the corresponding fuzzy matrices. Such approach can be very useful for certain situations in the area of fuzzy decision making. The following example may be taken as an illustration for the same.

Example 2.6.2. (Lin and Lee [45]). Let in a department four courses, say C_1, C_2, C_3 and C_4 be offered. Three students, say S_1, S_2 and S_3 are planning to take one of these courses based on different preferences for certain properties, say theory (T), application (A), hardware (H) and programming ($\mathcal{P}$). Let $X = (S_1, S_2, S_3)$, $Y = (T, A, H, P)$ and $Z = (C_1, C_2, C_3, C_4)$. Let the following data be given.

(i) The student's interest for various attributes T, A, H and $\mathcal{P}$ is represented by binary fuzzy relation

$$\mathcal{R}(X,Y) = \begin{array}{c} \\ S_1 \\ S_2 \\ S_3 \end{array} \begin{array}{c} \begin{array}{cccc} T & A & H & \mathcal{P} \end{array} \\ \begin{pmatrix} 0.2 & 1.0 & 0.8 & 0.1 \\ 1.0 & 0.1 & 0.0 & 0.5 \\ 0.5 & 0.9 & 0.5 & 1.0 \end{pmatrix} \end{array}.$$

(ii) The properties of the courses C_1, C_2, C_3 and C_4 are indicated by the fuzzy relation $\mathcal{Q}(Y,Z)$ where

$$\mathcal{Q}(Y,Z) = \begin{array}{c} \\ T \\ A \\ H \\ \mathcal{P} \end{array} \begin{array}{c} \begin{array}{cccc} C_1 & C_2 & C_3 & C_4 \end{array} \\ \begin{pmatrix} 1.0 & 0.5 & 0.6 & 0.1 \\ 0.2 & 1.0 & 0.8 & 0.8 \\ 0.0 & 0.3 & 0.7 & 0.0 \\ 0.1 & 0.5 & 0.8 & 1.0 \end{pmatrix} \end{array}.$$

Then one may compute the standard max-min composition of $\mathcal{R}(X,Y)$ and $\mathcal{Q}(Y,Z)$ as follows

$$\mathcal{R} \circ \mathcal{Q} = \begin{pmatrix} 0.2 & 1.0 & 0.8 & 0.1 \\ 1.0 & 0.1 & 0.0 & 0.5 \\ 0.5 & 0.9 & 0.5 & 1.0 \end{pmatrix} \circ \begin{pmatrix} 1.0 & 0.5 & 0.6 & 0.1 \\ 0.2 & 1.0 & 0.8 & 0.8 \\ 0.0 & 0.3 & 0.7 & 0.0 \\ 0.1 & 0.5 & 0.8 & 1.0 \end{pmatrix}$$

$$= \begin{array}{c} \\ S_1 \\ S_2 \\ S_3 \end{array} \begin{array}{c} \begin{array}{cccc} C_1 & C_2 & C_3 & C_4 \end{array} \\ \begin{pmatrix} 0.2 & 1.0 & 0.8 & 0.8 \\ 1.0 & 0.5 & 0.6 & 0.5 \\ 0.5 & 0.9 & 0.8 & 1.0. \end{pmatrix} \end{array}.$$

Here the $(1,1)^{\text{th}}$ element of $\mathcal{R} \circ Q$ is obtained as $(0.2 \wedge 1.0) \vee (1.0 \wedge 0.2)$ $\vee(0.8 \wedge 0.0) \vee (0.8 \wedge 0.1)$ i.e. 0.2. Similarly the other elements of $\mathcal{R} \circ Q$ are computed.

The fuzzy matrix for the composition $\mathcal{R} \circ Q$ indicates that the student S_1 should be advised to take the course C_2, S_2 the course C_1 and S_3 the course C_4.

Remark 2.6.3. Once a binary fuzzy relation $\mathcal{R}(X, X)$ on X is defined, one can define $\mathcal{R}$ to be *reflexive* if $\mu_{\mathcal{R}}(x, x) = 1$ for all $x \in X$. Further $\mathcal{R}$ can be called *symmetric* if $\mu_{\mathcal{R}}(x, y) = \mu_{\mathcal{R}}(y, x)$ for all $x,\ y \in X$ and it can be called *transitive* if $\mu_{\mathcal{R}}(x, z) \geq \max_y \min \big(\mu_{\mathcal{R}}(x, y),\ \mu_{\mathcal{R}}(y, z)\big)$ for all $x,\ z \in X$. This will eventually lead to the fuzzy analogue of an equivalence relation and that in the fuzzy context is called as a *similarity relation.* Thus a relation $\mathcal{R}(X, X)$ on X is a similarity relation if it is reflexive, symmetric and transitive. If $\mathcal{R}(X, X)$ is only reflexive and symmetric then it is called a *resemblance relation.* Further one can define $\mathcal{R}(X, X)$ to be *antisymmetric* if $\mu_{\mathcal{R}}(x, y) \neq \mu_{\mathcal{R}}(y, x)$ for all elements in the support of the relation $\mathcal{R}$. A binary fuzzy relation $\mathcal{R}$ on X that is reflexive, antisymmetric and transitive is called a *fuzzy partial ordering.* Though such studies are extremely useful in the areas of fuzzy databases and fuzzy pattern recognition theory, they will not be pursued here and one can refer to the texts already mentioned in the introduction.

2.7 Triangular norms (*t*-norms) and triangular conorms (*t*-conorms)

While discussing the standard union, standard intersection and standard complement of fuzzy sets, it was noted in Section 2.2 that because of the unsharp boundary, a fuzzy set A in X and its standard complement A' overlap and therefore do not satisfy the law of excluded middle and law of contradiction, i.e. $A \cup A' = X$ and $A \cap A' = \emptyset$ do not hold. As it happens, the standard operations of union, intersection and complement for fuzzy sets in X are not the only possible generalization of corresponding crisp set operations. It is possible to define these operations for fuzzy sets in a more general way by using the concept of triangular norms (t-norms) and triangular conorms (t-conorms), which have been developed by Schweizer and Sklar [69] in the context of statistical metric spaces. Although there is a large amount of literature on t-norms and t-conorms, the presentation here is very brief and for a more detailed account one may refer to texts like Dubois and Prade [15] and Dumitrescu, Lazzerini and Jain [17].

Definition 2.7.1 (t-norm). *A function $T : [0,1] \times [0,1] \to [0,1]$ is called a t-norm if it satisfies the following axioms*

(i) $T(a,1) = a$ *for all* $a \in [0,1]$ *(boundary condition),*
(ii) $T(a,b) \le T(u,v)$ *for* $a \le u,\ b \le v$ *(monotonicity),*
(iii) $T(a,b) = T(b,a)$ *(commutativity),*
(iv) $T\big(T(a,b),c\big) = T\big(a,T(b,c)\big)$ *(associativity).*

Definition 2.7.2 (Archimedean t-norm). *A t-norm T is said to Archimedean if* $T(a,a) < a$ *for all* $a \in (0,1)$.

Remark 2.7.1. From the axioms of t-norm it can be proved that $T(0,a) = T(a,0) = 0$ and also $T(a,a) \le T(a,1) = a$ for all $a \in [0,1]$. Therefore the t-norm T is Archimedean if and only if there does not exist any $a \in (0,1)$ with $T(a,a) = a$.

Definition 2.7.3 (t-Conorm). *A function $S : [0,1] \times [0,1] \to [0,1]$ which is commutative, associative and monotonic in every variable with* $S(a,0) = a$ *for all* $a \in [0,1]$, *is called a triangular conorm or t-conorm.*

Definition 2.7.4 (Archimedean t-conorm). *A t-conorm S is said to be Archimedean if* $S(a,a) > a$ *for all* $a \in (0,1)$.

Let T be a t-norm. The function $S : [0,1] \times [0,1] \to [0,1]$ defined by $S(a,b) = 1 - T(1-a, 1-b)$ for all $a, b \in [0,1]$ is called a t-conorm or the *dual* of t-norm T.

Remark 2.7.2. Using the axioms of t-norm and the definition of its dual, it can be proved that S is monotone, commutative and associative. Also $S(a,0) = a,\ S(a,1) = 1$ and $S(a,a) \ge a$ for all $a \in [0,1]$. Further if T is an Archimedean t-norm then so is its dual t-conorm. Some of the most common and useful t-norms and t-conorms are following

(i) $T_0(a,b) = \min(a,\ b),\ S_0(a,b) = \max(a,\ b)$,

(ii) $T_1(a,b) = ab,\ S_1(a,b) = a + b - ab,$
(iii) $T_\infty(a,b) = \max\,(a + b - 1,\ 0),\ S_\infty(a,b) = \min\,(a + b,\ 1).$

Here it can be verified that T_0 is not Archimedean but T_1 and T_∞ are Archimedean. There are various other families of t-norms and t-conorms (e.g. Frank's fundamental family, Yager's family, Hamacher's family, Schweizer and Sklar's family, Sugeno's family and Dubois and Prade's family), but these are not discussed here. In fact $(T_0,\ S_0)$, $(T_1,\ S_1)$ and $(T_\infty,\ S_\infty)$ are particular members of Frank's fundamental family $(T_s,\ S_s)$, $0 \leq s \leq \infty$. For these and other details one may refer to Dumitrescu, Lazzerini and Jain [17].

As t-norms and t-conorms are more general connectives than the usual *min* and *max* operators, one can define operations on fuzzy sets using these triangular norms and conorms.

Definition 2.7.5 (Intersection of fuzzy sets). *Let A and B be two fuzzy sets in X. The intersection of A and B with respect to a given t-norm T is defined as a fuzzy set $A \cap B$ whose membership function is given by*

$$\mu_{A\cap B}(x) = T\big(\mu_A(x),\ \mu_B(x)\big),\ x \in X.$$

Definition 2.7.6 (Union of fuzzy sets). *Let A and B be two fuzzy sets in X. The union of A and B with respect to a given conorm S is defined as a fuzzy set $A \cup B$ whose membership function is given by*

$$\mu_{A\cup B}(x) = S\big(\mu_A(x),\ \mu_B(x)\big),\ x \in X.$$

Here it may be noted that in the above definitions, T and S are not necessarily dual norms. Also the standard union and intersection of fuzzy sets using *min* and *max* operations are special cases of above definitions for the pair $(T_0,\ S_0)$. The pair $(T_0,\ S_0)$ still remains popular and useful in the theory of fuzzy sets because of the following theorem due to Bellman and Giertz [8] whose proof is not given here.

Theorem 2.7.1 *Let $F,\ G : [0,1] \times [0,1] \to [0,1]$. Then $F = T_0$ and $G = S_0$ are the only functions that satisfy the following:*

(i) F and G are commutative,
(ii) F and G are associative,
(iii) F and G are mutually distributive,
(iv) $F(x,y) \leq T_0(x,y)$, $G(x,y) \geq S_0(x,y)$ for all $x,\ y \in [0,1]$,
(v) $F(1,1) = 1$, $G(0,0) = 0$,

(vi) F and G are continuous,
(vii) $F(a,b) \leq F(u,v)$ and $G(a,b) \leq G(u,v)$ for all $a \leq u$, $b \leq v$,
(viii) $F(x,x) < F(y,y)$ and $G(x,x) < G(y,y)$ if $x < y$.

Definition 2.7.7 (Complement function). *A complement function is a function $C : [0,1] \rightarrow [0,1]$ if*

(i) $C(0) = 1$, $C(1) = 0$,
(ii) $C \circ C(a) = a$ for all $a \in [0,1]$,
(iii) C is strictly decreasing and
(iv) C is continuous.

Definition 2.7.8 (Fuzzy complement). *Let A be a fuzzy set in X. The complement of the fuzzy set A with respect to a complement function C is a fuzzy set A' whose membership function is given by*

$$\mu_{A'}(x) = C\big(\mu_A(x)\big),\ x \in X.$$

Remark 2.7.3. The standard complement of a fuzzy set A in X is obtained when the complement function C is taken as the standard negation function i.e.

$$N(a) = 1 - a \text{ for all } a \in [0,1].$$

Definition 2.7.9 (C-dual). *Let T and S be a t-norm and t-conorm respectively. The pair (T, S) is said to be dual with respect to a complement function C or C-dual if*

$$T\big(C(a), C(b)\big) = C\big(S(a,b)\big),\ a, b \in [0,1].$$

Remark 2.7.4. (i) Let N be the standard negation. Then T and S are N-dual if and only if S is the dual conorm of T i.e.

$$S(a,b) = 1 - T(1-a, 1-b),\ a,\ b \in [0,1].$$

(ii) It can be proved that for any t-norm T and any complement function C, the t-conorm S defined by

$$S(a,b) = C\big(T(C(a),\ C(b))\big),\ a,\ b \in [0,1],$$

gives a C-dual conorm of T.

(iii) The notion of C-duality is nothing but a generalization of standard De Morgan law $(A \cup B)' = A' \cap B'$.

Remark 2.7.5. Taking the standard intersection, union and complementation, i.e. with respect to the pair $(T_0,\ S_0)$ and the standard negation N, it has been observed earlier that the excluded middle law $(A \cup A' = X)$ and law of contradiction $(A \cap A' = \emptyset)$ do not hold for the fuzzy sets. What is interesting in this context that no matter which complement function C is taken, the fuzzy set operations induced by the pair $(T_0,\ S_0)$ do not satisfy these laws with respect to C.

Remark 2.7.6. For the pair $(T_\infty,\ S_\infty)$ the intersection and union of two fuzzy set A and B in X are defined by

$$\mu_{A\cap B}(x) = \max\ \Big(\mu_A(x) + \mu_B(x) - 1,\ 0\Big)$$
$$\mu_{A\cup B}(x) = \min\ \Big(\mu_A(x) + \mu_B(x),\ 1\Big)$$

for $x \in X$. Here it can be proved that for the fuzzy set operations induced by the pair $(T_\infty,\ S_\infty)$, the laws of excluded middle and contradiction hold if and only if fuzzy complement is induced by the standard negation. However for the pair $(T_\infty,\ S_\infty)$, the intersection and union are not idempotent i.e. $A \cup A \neq A$ and $A \cap A \neq A$.

Further for the pair $(T_\infty,\ S_\infty)$ and the complementation given by the standard negation function, the intersection and union are not mutually distributive. This later result holds in somewhat more generality as stated in the below given theorem.

Theorem 2.7.2 *Let the pair* $(T,\ S)$ *be the* C*-dual. If for the fuzzy set operations induced by the pair* $(T,\ S)$ *the laws of excluded middle and contradiction hold with respect to the complement function* C*, then the intersection and union are not mutually distributive.*

For the proof of Theorem 2.7.2 and various results mentioned in Remarks 2.7.5 and 2.7.6, one may refer Dumitrescu, Lazzzerini and Jain [17].

Remark 2.7.7. In this section, it has been emphasized that contrary to the crisp scenario, the fuzzy set theoretic operations of intersection, union and complementation are not uniquely defined. These operations depend on the specific choice of the pair $(T,\ S)$ and the complement function C. Therefore membership function of fuzzy sets as well as operations on fuzzy sets depend on the particular context in which the specific problem is being studied. For a meaningful application of fuzzy set theory to real life problems the correct subjective choice of membership functions and a meaningful choice of the pair $(T,\ S)$ is important.

2.8 Conclusions

In this chapter we have presented certain very basic definitions and results on fuzzy sets and related topics. There are many more mathematical generalizations of fuzzy sets but from the applications point of view the notion of type-2 fuzzy sets is rather interesting. The type-2 fuzzy sets have recently been applied to areas like AI, forecasting of time series, knowledge-mining, and digital communications etc. An appropriate reference for type-2 fuzzy sets is the book by Mandel [51].

3

Fuzzy numbers and fuzzy arithmetic

3.1 Introduction

While modeling certain problems in the physical sciences and engineering, it is often observed that the parameters of the problem are not known precisely but rather lie in an interval. In the past, such situations have been handled by the application of *interval arithmetic* (Moore [55], [56]) which allows mathematical computations (operations) to be performed on intervals and obtain meaningful estimates of desired quantities also in terms of intervals. *Fuzzy arithmetic* (arithmetic of *fuzzy numbers*) can be taken as a generalization of the interval arithmetic where rather than considering intervals at one (constant) level only, several levels in [0,1] are considered. This is primarily because of the basic definition of a fuzzy set which allows gradation of membership for an element of the universal set. Because of this, the modeling based on fuzzy arithmetic is expected to express the situation more realistically.

This chapter is divided into five main sections, namely, *interval arithmetic, fuzzy numbers and their representations, arithmetic of fuzzy numbers, special types of fuzzy numbers and their arithmetic* and *ranking of fuzzy numbers.* Some appropriate references for this Chapter are Dubois and Prade ([14], [15]) and Kaufmann and Gupta ([32], [33]).

3.2 Interval arithmetic

To understand the fundamentals of fuzzy arithmetic one needs to learn about interval arithmetic, i.e given two closed intervals in $\mathbb{R}$, how to "add", "subtract","multiply" and "divide" these intervals. In this context, a closed interval in $\mathbb{R}$ is also called an *interval of confidence* as it limits the uncertainty of data to an interval. Let $A = [a_1, a_2]$ and $B = [b_1, b_2]$ be two closed intervals in $\mathbb{R}$. Then we have the following definitions:

Definition 3.2.1 (Addition (+) and subtraction (−)). *If $x \in [a_1, a_2]$, $y \in [b_1, b_2]$ then $x + y \in [a_1 + b_1, a_2 + b_2]$ and $x - y \in [a_1 - b_2, a_2 - b_1]$. Therefore the addition of A and B, denoted by $A(+)B$, is defined as*

$$A(+)B = [a_1, a_2](+)[b_1, b_2] = [a_1 + b_1, a_2 + b_2].$$

Similarly, the subtraction of A and B, denoted by $A(-)B$ is defined as

$$A(-)B = [a_1, a_2](-)[b_1, b_2] = [a_1 - b_2, a_2 - b_1].$$

Definition 3.2.2 (Image of an interval). *If $x \in [a_1, a_2]$ then its image $-x \in [-a_2, -a_1]$. Therefore the image of A, denoted by $\overline{A}$ is defined as*

$$\overline{A} = \overline{[a_1, a_2]} = [-a_2, -a_1].$$

Definition 3.2.3 (Multiplication (·)). *The multiplication of two closed intervals $A = [a_1, a_2]$ and $B = [b_1, b_2]$ of $\mathbb{R}$, denoted by $A(\cdot)B$, is defined as*

$$\begin{aligned} A(\cdot)B &= [a_1, a_2](\cdot)[b_1, b_2] \\ &= \Big[\min(a_1b_1, a_1b_2, a_2b_1, a_2b_2), \max(a_1b_1, a_1b_2, a_2b_1, a_2b_2)\Big]. \end{aligned}$$

In case these intervals are in $\mathbb{R}_+$, the non-negative real line, the multiplication formula gets simplified to

$$A(\cdot)B = [a_1b_1, a_2b_2].$$

Definition 3.2.4 (Scalar multiplication and inverse). *Let $A = [a_1, a_2]$ be a closed interval in $\mathbb{R}_+$ and $k \in \mathbb{R}_+$. Identifying the scalar k as the closed interval $[k, k]$, the scalar multiplication $k \cdot A$ is defined as*

$$k \cdot A = [k, k](\cdot)[a_1, a_2] = [ka_1, ka_2].$$

Also, for $A = [a_1, a_2]$ in $\mathbb{R}_+$, if $x \in [a_1, a_2]$ and $0 \notin [a_1, a_2]$ then $\left(\frac{1}{x}\right) \in \left[\frac{1}{a_2}, \frac{1}{a_1}\right]$. Therefore the inverse of A, denoted by A^{-1}, is defined as

$$A^{-1} = [a_1, a_2]^{-1} = \left[\frac{1}{a_2}, \frac{1}{a_1}\right],$$

provided $0 \notin [a_1, a_2]$.

Definition 3.2.5 (Division (:)). *The division of two closed intervals* $A = [a_1, a_2]$ *and* $B = [b_1, b_2]$ *of* $\mathbb{R}$, *denoted by* $A(:)B$ *is defined as the multiplication of* $[a_1, a_2]$ *and* $\left[\frac{1}{b_2}, \frac{1}{b_1}\right]$ *provided* $0 \notin [b_1, b_2]$. *Therefore*

$$\begin{aligned} A(:)B &= [a_1, a_2](:)[b_1, b_2] \\ &= [a_1, a_2](\cdot)\left[\frac{1}{b_2}, \frac{1}{b_1}\right] \\ &= \left[\min\left(\frac{a_1}{b_2}, \frac{a_1}{b_1}, \frac{a_2}{b_2}, \frac{a_2}{b_1}\right), \max\left(\frac{a_1}{b_2}, \frac{a_1}{b_1}, \frac{a_2}{b_2}, \frac{a_2}{b_1}\right)\right]. \end{aligned}$$

In case these intervals are in $\mathbb{R}_+$ and as before $0 \notin [b_1, b_2]$, this formula for the division gets simplified to

$$A(:)B = \left[\frac{a_1}{b_2}, \frac{a_2}{b_1}\right].$$

Further, one can identify $A(:)B \equiv A(\cdot)B^{-1}$ provided $0 \notin B = [b_1, b_2]$ and obtain the same formula as given earlier. Along the lines of scalar multiplication, the *division by a scalar* $k > 0$ can also be defined as

$$A(:)k = [a_1, a_2](\cdot)\left[\frac{1}{k}, \frac{1}{k}\right] = \left[\frac{a_1}{k}, \frac{a_2}{k}\right].$$

Definition 3.2.6 (Max ($\vee$) and min ($\wedge$) operations). *Let* $A = [a_1, a_2]$ *and* $B = [b_1, b_2]$ *be two closed intervals in* $\mathbb{R}$. *Then the* max ($\vee$) *and* min ($\wedge$) *operations on A and B are defined as*

$$\begin{aligned} A(\vee)B &= [a_1, a_2](\vee)[b_1, b_2] = [a_1 \vee b_1, a_2 \vee b_2], \\ A(\wedge)B &= [a_1, a_2](\wedge)[b_1, b_2] = [a_1 \wedge b_1, a_2 \wedge b_2]. \end{aligned}$$

Remark 3.2.1. It can be verified that addition (+) and multiplication ($\cdot$) operations on closed intervals as defined above are commutative and associative but subtraction (−) and division (:) are neither commutative nor associative. Also, $A(+)\overline{A} = [a_1, a_2](+)[-a_2, -a_1] \neq [0, 0] \equiv 0$. In case $A = [a_1, a_2]$ is in $\mathbb{R}_+$ and $0 \notin [a_1, a_2]$, $A(\cdot)A^{-1} = A^{-1}(\cdot)A \neq [1, 1] \equiv 1$.

We now proceed to the next section to introduce fuzzy numbers and their associated arithmetic. As it turns out, the fuzzy arithmetic is essentially the arithmetic of α-cuts A_α which are given by closed and bounded intervals of type $A_\alpha = [a_\alpha^L, a_\alpha^R]$, $\alpha \in (0,1]$, when the fuzzy set A in $\mathbb{R}$ is a *fuzzy number*. Therefore all the ideas presented in this section can be borrowed for intervals of type $[a_\alpha^L, a_\alpha^R]$, $\alpha \in (0,1]$ and a meaningful arithmetic of fuzzy numbers be developed.

3.3 Fuzzy numbers and their representation

There are many real life situations, in areas like decision making and optimization, where rather than dealing with crisp real numbers and crisp intervals, one has to deal with "approximate" numbers or intervals of type "number that are close to a given real number" or "numbers that are around a given interval of real number". The purpose of this section is to understand that how such fuzzy statements can be conceptualized by certain "appropriate" fuzzy sets in $\mathbb{R}$ to be termed as *fuzzy numbers*.

For the motivation to define a fuzzy number, let us consider the fuzzy statement "numbers that are close to a given real number r". Since the real number r is certainly close to r itself, any fuzzy set A in $\mathbb{R}$ which tries to represent the given fuzzy statement must have the property that $\mu_A(r) = 1$, i.e. A must be a normal fuzzy set. Also, just prescribing an interval around r is not enough. The intervals should be considered at varying levels $\alpha \in (0,1]$ to have the proper gradation i.e. the α-cuts of A must be closed intervals of the type $[a_\alpha^L, a_\alpha^R]$. Further, to carry out interval arithmetic as described in the previous section, the intervals $[a_\alpha^L, a_\alpha^R]$ for $\alpha \in (0,1]$ must be of finite length and for that one needs that the support of A is bounded. Therefore it makes sense to define a fuzzy number as follows.

Definition 3.3.1 (Fuzzy number). *A fuzzy set A in $\mathbb{R}$ is called a fuzzy number if it satisfies the following conditions*

(i) A is normal,
(ii) A_α is a closed interval for every $\alpha \in (0,1]$,
(iii) the support of A is bounded.

The theorem presented below gives a complete characterization of a fuzzy number.

Theorem 3.3.1 *Let A be a fuzzy set in $\mathbb{R}$. Then A is a fuzzy number if and only if there exists a closed interval (which may be singleton) $[a, b] \neq \phi$ such that*

$$\mu_A(x) = \begin{cases} 1, & x \in [a, b], \\ l(x), & x \in (-\infty,\ a), \\ r(x), & x \in (b,\ \infty), \end{cases}$$

where (i) $l : (-\infty,\ a) \to [0,1]$ *is increasing, continuous from the right and* $l(x) = 0$ *for* $x \in (-\infty,\ w_1)$, $w_1 < a$ *and (ii)* $r : (b,\ \infty) \to [0,1]$ *is decreasing continuous from the left and* $r(x) = 0$ *for* $x \in (w_2,\ \infty), w_2 > b$.

In the above theorem the term "increasing" is to be understood in the sense that "$x \geq y \Longrightarrow l(x) \geq l(y)$" i.e. l is non-decreasing. Although the proof of this theorem is not given here. but one can refer to Klir and Yuan [35] for the detailed proof. Its consequence is of special significance which tells that if a fuzzy set A in $\mathbb{R}$ represents a fuzzy number, how will its membership functions μ_A will look like.

Remark 3.3.1. In case the membership function of the fuzzy set A in $\mathbb{R}$ takes the form $\mu_A(x) = 1$ for $x = a$ and $\mu_A(x) = 0$ for $x \neq a$, it becomes the characteristic function of the singleton set $\{a\}$ and therefore represents the real number a. A real interval $[a,\ b]$ can also be identified similarly by its characteristic function. In most of the practical applications the function $l(x)$ and $r(x)$ are continuous which give the continuity of the membership function. The following figures are self explainary.

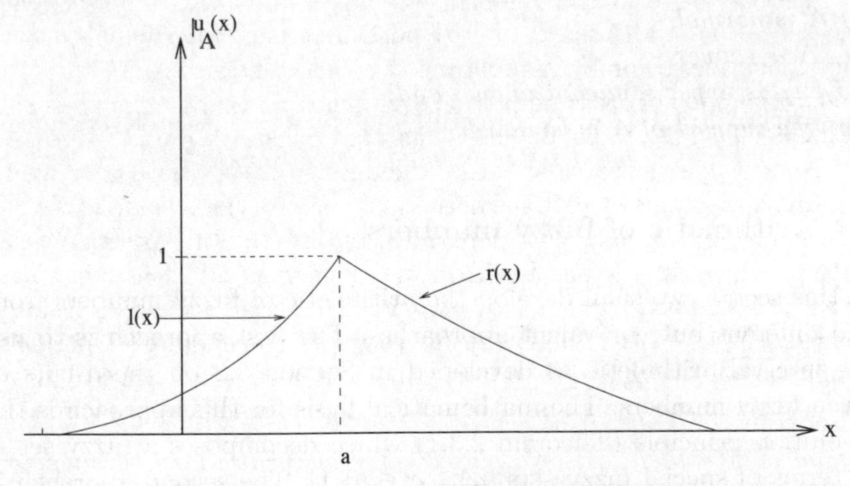

Fig. 3.1. Fuzzy number a, with continuous l and r.

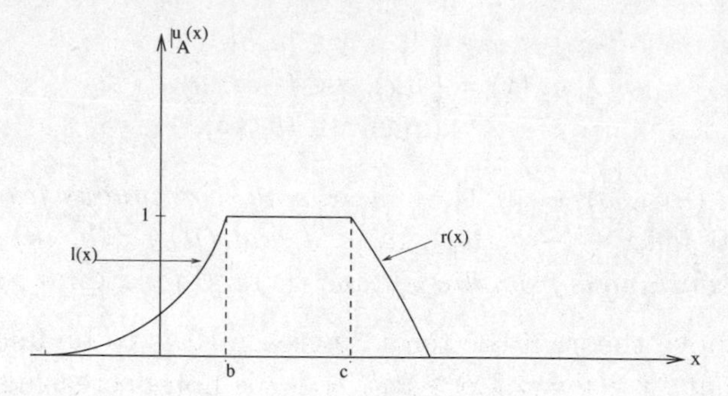

Fig. 3.2. Fuzzy interval $[b, c]$ with continuous l and r.

Remark 3.3.2. It may be recalled that a fuzzy set A in X is a convex fuzzy set if and only if all its α-cuts A_α are convex (crisp) sets for $\alpha \in (0, 1]$. Here in the context of fuzzy numbers, $X \equiv \mathbb{R}$ and it is known that the only convex sets in $\mathbb{R}$ are intervals. Further when the membership function of the convex fuzzy set A in $\mathbb{R}$ is upper semicontinuous, then all these α-cuts A_α, for $\alpha \in (0, 1]$, are closed intervals. Since the basic requirement to define a fuzzy number is that all its α-cuts A_α, $\alpha \in (0, 1]$, are closed and bounded intervals, we may also have the following alternate definition of a fuzzy number as per Definition 3.3.2 given below.

Definition 3.3.2 (Fuzzy number). *Let A be a fuzzy set in $\mathbb{R}$. then A is called a fuzzy number if*

(i) A is normal,
(ii) A is convex,
(iii) μ_A is upper semicontinuous, and,
(iv) the support of A is bounded.

3.4 Arithmetic of fuzzy numbers

In this section we shall develop the arithmetic of fuzzy numbers from two different but equivalent approaches. The first approach is to use the interval arithmetic as developed in Section 3.2 on the α-cuts of given fuzzy numbers. The mathematical basis for this approach is the resolution principle (Theorem 2.3.1) which decomposes a fuzzy set A in terms of special fuzzy sets αA_α, $\alpha \in (0, 1]$. The second approach is based on the extension principle of Zadeh which has been presented in Section 2.5.

Approach based on α-cuts

Let A and B be two fuzzy numbers and $A_\alpha = [a_\alpha^L, a_\alpha^R]$, $B_\alpha = [b_\alpha^L, b_\alpha^R]$ be α-cuts, $\alpha \in (0,1]$, of A and B respectively. Let $*$ denote any of the arithmetic operations $(+)$, $(-)$, $(\cdot)$, $(:)$, $\wedge$, $\vee$ on fuzzy numbers. Then we have the following definition.

Definition 3.4.1 ($*$ Operation on two fuzzy numbers). *Let A, B, A_α and B_α be as described above. Then the $*$ operation on fuzzy numbers A and B, denoted by $A * B$, gives a fuzzy number in $\mathbb{R}$ where*

$$A * B = \bigcup_\alpha \alpha (A * B)_\alpha,$$

and

$$(A * B)_\alpha = A_\alpha * B_\alpha, \quad \alpha \in (0,1].$$

Here it may be remarked that the reason for $A * B$ to be a fuzzy number, and not just a general fuzzy set, is that A and B being fuzzy numbers, the sets A_α, B_α, $(A * B)_\alpha$ are all closed intervals for $\alpha \in (0,1]$. Also for a given $\alpha \in (0,1]$, the closed interval $(A * B)_\alpha$ can be computed by applying the interval arithmetic on the closed intervals A_α and B_α with respect to the operation $*$. In particular,

$$A_\alpha(+)B_\alpha = [a_\alpha^L + b_\alpha^L,\ a_\alpha^R + b_\alpha^R],$$
$$A_\alpha(-)B_\alpha = [a_\alpha^L - b_\alpha^R,\ a_\alpha^R - b_\alpha^L].$$

Further for fuzzy numbers A and B in $\mathbb{R}_+$,

$$A_\alpha(\cdot)B_\alpha = [a_\alpha^L b_\alpha^L,\ a_\alpha^R b_\alpha^R],$$
$$A_\alpha(:)B_\alpha = \left[\frac{a_\alpha^L}{b_\alpha^R}, \frac{a_\alpha^R}{b_\alpha^L}\right], \ 0 \notin [b_\alpha^L, b_\alpha^R].$$

The multiplication of a fuzzy number A in $\mathbb{R}$ by a real number $k > 0$ can again be defined as earlier in the context of interval arithmetic, i.e.

$$(k \cdot A)_\alpha = k \cdot A_\alpha = [k a_\alpha^L,\ k a_\alpha^R].$$

Approach based on the Zadeh's extension principle

Let A and B be two fuzzy numbers and $*$ be any of the arithmetic operations described above. Then by using the Zadeh's extension principle, the fuzzy number $A * B$ is defined as

$$\mu_{A*B}(z) = \sup_{z=x*y} \min \left(\mu_A(x),\ \mu_B(y)\right), \text{ for all } z \in R.$$

In particular we have

$$\begin{aligned}
\mu_{A(+)B}(z) &= \sup_{z=x+y} \min \left(\mu_A(x),\ \mu_B(y)\right),\\
\mu_{A(-)B}(z) &= \sup_{z=x-y} \min \left(\mu_A(x),\ \mu_B(y)\right),\\
\mu_{A(\cdot)B}(z) &= \sup_{z=xy} \min \left(\mu_A(x),\ \mu_B(y)\right),\\
\mu_{A(:)B}(z) &= \sup_{z=x/y} \min \left(\mu_A(x),\ \mu_B(y)\right).
\end{aligned}$$

The arithmetic of fuzzy numbers as discussed above will be studied again in the next section where certain special types of fuzzy numbers and their arithmetic will be studied. As we shall see there, some of the arithmetical formulas presented here will get further simplified because of the special features of fuzzy numbers at hand. It is planned to present several examples in the next section to make this point more explicit and clear.

3.5 Special types of fuzzy numbers and their arithmetic

As the set of fuzzy numbers is rather large (uncountably infinite) and their arithmetic is in general computationally expensive, it is imperative to define and select a few special types of fuzzy numbers to be used for real life applications. Some such special types of fuzzy numbers and their arithmetic is being discussed here which will be used extensively in later chapters on fuzzy mathematical programming and fuzzy games.

Definition 3.5.1 (Triangular fuzzy number (TFN)). *A fuzzy number A is called a triangular fuzzy number (TFN) if its membership function μ_A is given by*

$$\mu_A(x) = \begin{cases} 0 & , x < a_l, x > a_u, \\ \dfrac{x - a_l}{a - a_l} & , a_l \le x \le a, \\ \dfrac{a_u - x}{a_u - a} & , a < x \le a_u. \end{cases}$$

The TFN A is denoted by the triplet $A = (a_l,\ a,\ a_u)$ and has the shape of a triangle as shown in the Figure 3.3.

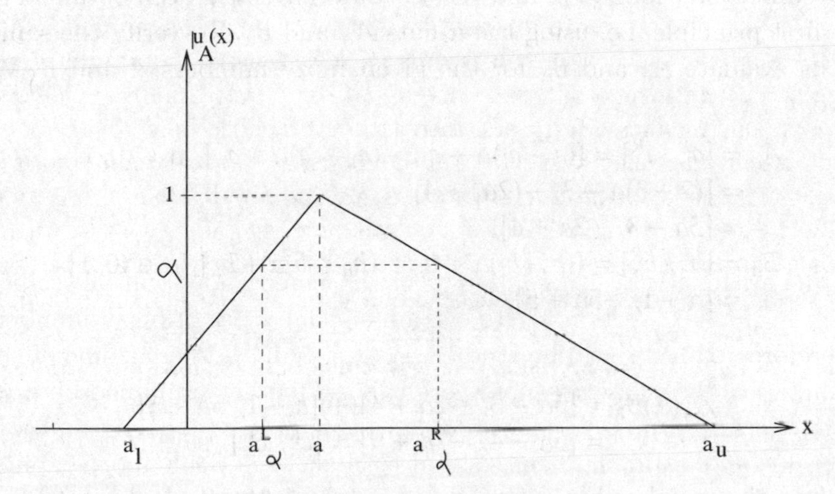

Fig. 3.3. A Triangular fuzzy number $A = (a_l,\ a,\ a_u)$

Further the α-cut of the TFN $A = [a_l,\ a,\ a_u]$ is the closed interval

$$A_\alpha = [a^L_\alpha,\ a^R_\alpha] = \Big[(a - a_l)\alpha + a_l,\ -(a_u - a)\alpha + a_u\Big],\ \ \alpha \in (0, 1].$$

Next let $A = (a_l,\ a,\ a_u)$ and $B = (b_l,\ b,\ b_u)$ be two TFNs then using the α-cuts, A_α and B_α for $\alpha \in (0, 1]$ one can compute $A * B$ where $*$ may be $(+)$, $(-)$, $(\cdot)$, $(:)$, $\vee$, $\wedge$ operation. In this context it can be verified that

$$A(+)B = (a_l + b_l,\ a + b,\ a_u + b_u),$$
$$-A = (-a_u,\ -a,\ -a_l),$$
$$kA = (ka_l,\ ka,\ ka_u), k > 0,$$

and,

$$A(-)B = (a_l - b_u,\ a - b,\ a_u - b_l)$$

are TFNs but A^{-1}, $A(\cdot)B$, $A(:)B$, $(A \vee B)$, $(A \wedge B)$ need not be a TFN. We now consider the following examples.

Example 3.5.1. (Kaufmann and Gupta [32]) Let $A = (-3,2,4)$ and $B = (-1,0,5)$ be two TFNs. Then using the formulas for the addition and subtraction of TFNs we get

$$A(+)B = (-3,2,4)(+)(-1,0,5) = (-4,2,9),$$

and

$$A(-)B = (-3,2,4)(-)(-1,0,5) = (-8,2,5).$$

The same result for $A(+)B$ and $A(-)B$ could have also been obtained by the first principle, i.e. using the α-cuts A_α and B_α. To verify the same, let us evaluate A_α and B_α for the given fuzzy numbers A and B. We have

$$\begin{aligned} A_\alpha &= [a_\alpha^L,\ a_\alpha^R] = [(a - a_l)\alpha + a_l,\ -(a_u - a)\alpha + a_u],\ \alpha \in (0,1] \\ &= [(2+3)\alpha - 3,\ -(2\alpha) + 4] \\ &= [5\alpha - 3,\ -2\alpha + 4], \\ B_\alpha &= [b_\alpha^L,\ b_\alpha^R] = [(b - b_l)\alpha + b_l,\ -(b_u - b)\alpha + b_u],\ \alpha \in (0,1] \\ &= [\alpha - 1,\ -5\alpha + 5]. \end{aligned}$$

Therefore

$$\begin{aligned} A_\alpha(+)B_\alpha &= [5\alpha - 3,\ -2\alpha + 4](+)[\alpha - 1,\ 5\alpha + 5] \\ &= [6\alpha - 4,\ -7\alpha + 9] = [c_\alpha^L,\ c_\alpha^R] \text{ (say)} . \end{aligned}$$

To find the membership function $\mu_{A(+)B}(x)$ of $A(+)B$, we have to find the range of $x \in \mathbb{R}$ where α-level sets are valid. Thus from c_α^L,

$$x = 6\alpha - 4 \Rightarrow \alpha = (x+4)/6,$$

and from c_α^R,

$$x = -7\alpha + 9 \Rightarrow \alpha = (9-x)/7.$$

Further α becomes 1 for $x = 2$. Therefore the membership function of $A(+)B$ comes out to be

$$\mu_{A(+)B}(x) = \begin{cases} 0 & , x < -4 \text{ or } x > 9, \\ \dfrac{(x+4)}{6} & , -4 \le x \le 2, \\ \dfrac{(-x+9)}{7} & , 2 < x \le 9, \end{cases}$$

which is nothing but the TFN (-4,2,9) as shown in the below given figure.

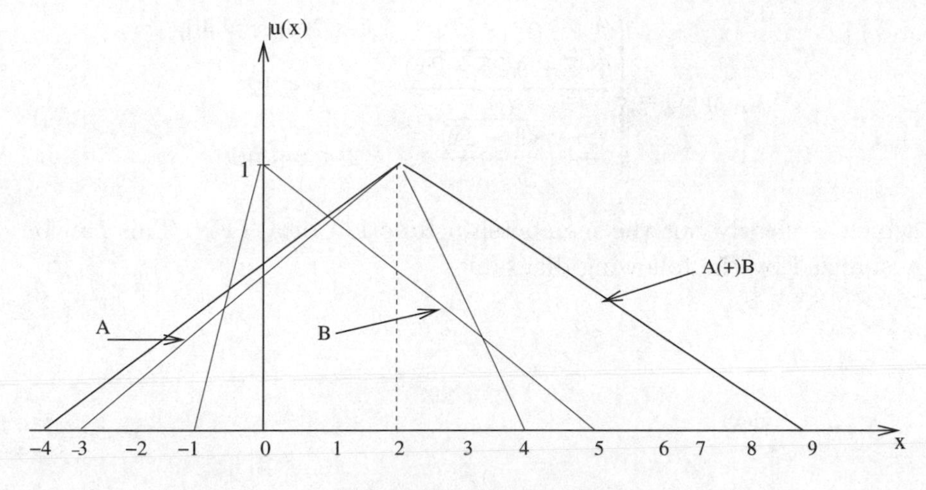

Fig. 3.4. A TFN $A(+)B$

Example 3.5.2. (Kaufmann and Gupta [32]) Let $A = (2, 3, 5)$ and $B = (1, 4, 8)$ be two TFNs in $\mathbb{R}_+$. As noted earlier and can also be seen in this example, the product $A(\cdot)B$ need not be a TFN in general, and therefore one has to compute $A(\cdot)B$ by the first principle, i.e. by using α-cuts A_α and B_α. Here we have

$$A_\alpha = [\alpha + 2, \ -2\alpha + 5], \ B_\alpha = [3\alpha + 1, \ -4\alpha + 8].$$
$$A_\alpha(\cdot)B_\alpha = [(\alpha + 2)(3\alpha + 1), (-2\alpha + 5)(-4\alpha + 8)]$$
$$= [3\alpha^2 + 7\alpha + 2, \ 8\alpha^2 - 36\alpha + 40] = [c_\alpha^L, \ c_\alpha^R].$$

For $\alpha = 0$, $A_0(\cdot)B_0 = [2, 40]$ and for $\alpha = 1$, $A_1(\cdot)B_1 = [12, 12] = 12$. Therefore the membership function of $A(\cdot)B$ takes the value 1 for $x = 12$ and is zero for $x < 2$ and also for $x > 40$. Also in between 2 and 12, and also between 12 and 40, the segments of the membership function are not straight lines, but parabola. Although it is intuitively clear from the presence of quadratic expressions in $A_\alpha(\cdot)B_\alpha$, one can obtain exact expressions for these parabolic segments. For this we need to find the range of $x \in \mathbb{R}_+$ where the α-level sets are valid. This can be accomplished from c_α^L by solving the equation $x = 3\alpha^2 + 7\alpha + 2$ i.e. $3\alpha^2 + 7\alpha + (2 - x) = 0$, i.e. $\alpha = (-7 \pm \sqrt{25 + 12x})/6$. Here only + sign will be taken because at $x = 12$, $\alpha = 1$ and α can not be negative. Similarly for c_α^R, one has to solve the equation $8\alpha^2 - 36\alpha + 40 = x$ i.e. $8\alpha^2 - 36\alpha + (40 - x) = 0$ and get $\alpha = (9 - \sqrt{1 - 2x})/5$. Therefore the membership function of $A(\cdot)B$ is given by

$$\mu_{A(\cdot)B}(x) = \begin{cases} 0 & , x < 2 \text{ or } x > 40, \\ \dfrac{(-7 + \sqrt{25 + 2x})}{6} & , 2 \le x \le 12, \\ \dfrac{(9 - \sqrt{1 - 2x})}{5} & , 12 < x \le 40, \end{cases}$$

which is clearly not the membership function of a TFN. This can be visualized by the following diagram

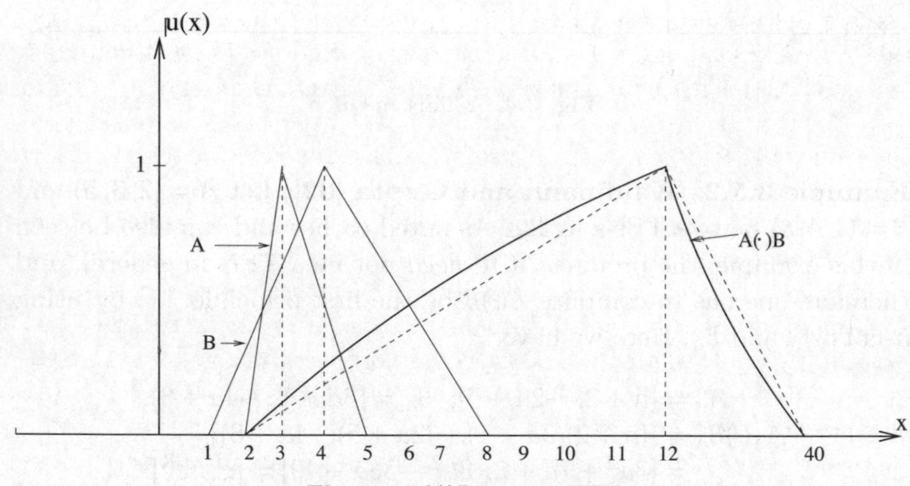

Fig. 3.5. $A(\cdot)B$ is not a TFN

Sometimes in practice, the fuzzy numbers $A(\cdot)B$, $A(:)B$, A^{-1}, $A \vee B$ and $A \wedge B$ which are not necessarily TFN can be approximated by a suitable TFN using the concept of left and right divergence. The TFN so obtained is called the *triangular approximation* of the given fuzzy number. A possible triangular approximation of $A(\cdot)B$ in our example here is the TFN (2,12,40). For more details on triangular approximation of fuzzy numbers one has to refer to Kauffmann and Gupta ([32], [33]) and Dubois and Prade ([14], [15]).

Definition 3.5.2 (Trapezoidal fuzzy number (TrFN)). *A fuzzy number A is called a trapezoidal fuzzy number if its membership function is given by*

$$\mu_A(x) = \begin{cases} 0 & , \ x < a_l, x > a_u, \\ \dfrac{x - a_l}{\underline{a} - a_l} & , \ a_l \le x < \underline{a}\,, \\ 1 & , \ \underline{a} \le x \le \overline{a}, \\ \dfrac{a_u - x}{a_u - \overline{a}} & , \ \overline{a} < x \le a_u. \end{cases}$$

The TrFN A is denoted by the quadruplet $A = (a_l,\ \underline{a},\ \overline{a},\ a_u)$ and has the shape of a trapezoid as shown in the figure 3.6.

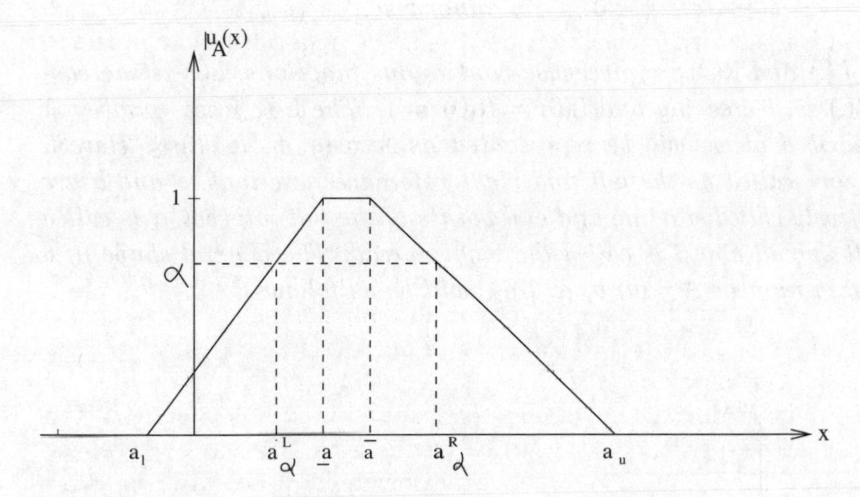

Fig. 3.6. A TrFN $A = (a_l,\ \underline{a},\ \overline{a},\ a_u)$

Further the α-cut of the TrFN $A = (a_l,\ \underline{a},\ \overline{a},\ a_u)$ is the closed interval $A_\alpha = [a^L_\alpha,\ a^R_\alpha] = [(\underline{a} - a_l)\alpha + a_l,\ -(a_u - \overline{a})\alpha + a_u],\ \alpha \in (0, 1]$.

Next let $A = (a_l,\ \underline{a},\ \overline{a},\ a_u)$ and $B = (b_l,\ \underline{b},\ \overline{b},\ b_u)$ be two TrFN, then using the α-cuts one can compute $A{*}B$ where $*$ may be $(+)$, $(-)$, $(\cdot)$, $(:)$, $\vee$ or $\wedge$ operation. In this context it can be verified that

$$A(+)B = (a_l + b_l,\ \underline{a} + \underline{b},\ \overline{a} + \overline{b},\ a_u + b_u),$$
$$-A = (-a_u,\ -\overline{a},\ -\underline{a},\ -a_l),$$
$$A(-)B = (a_l - b_u,\ \underline{a} - \overline{b},\ \overline{a} - \underline{b},\ a_u - b_l),$$

and,

$$kA = (ka_l,\ k\underline{a},\ k\overline{a},\ ka_u),\ k > 0,$$

are TrFNs but A^{-1}, $A(\cdot)B$, $A(:)B$, $A \vee B$ and $A \wedge B$ need not be a TrFN.

Definition 3.5.3 (*L-R* Fuzzy number). *A fuzzy number A is called a L-R fuzzy number if its membership function* $\mu_A : \mathbb{R} \to [0,1]$ *has the following form:*

$$\mu_A(x) = \begin{cases} L\left(\dfrac{x-a}{\alpha}\right), & (a-\alpha) \leq x < a,\ \alpha > 0, \\ 1, & a \leq x \leq b, \\ R\left(\dfrac{x-b}{\beta}\right), & b < x \leq (b+\beta),\ \beta > 0, \\ 0, & \text{otherwise,} \end{cases}$$

where L(.) and R(.) are piecewise continuous functions, L(.) is increasing, R(.) is decreasing and L(0) = R(0) = 1. The L-R fuzzy number A as described above will be represented as $A = (a,\ b,\ \alpha,\ \beta)_{LR}$. *Here L and R are called as the left and right reference functions, a and b are respectively called starting and end points of the flat interval, α is called the left spread and β is called the right spread. The general shape of a L-R fuzzy number* $A = (a,\ b,\ \alpha,\ \beta)_{LR}$ *will be as follows*

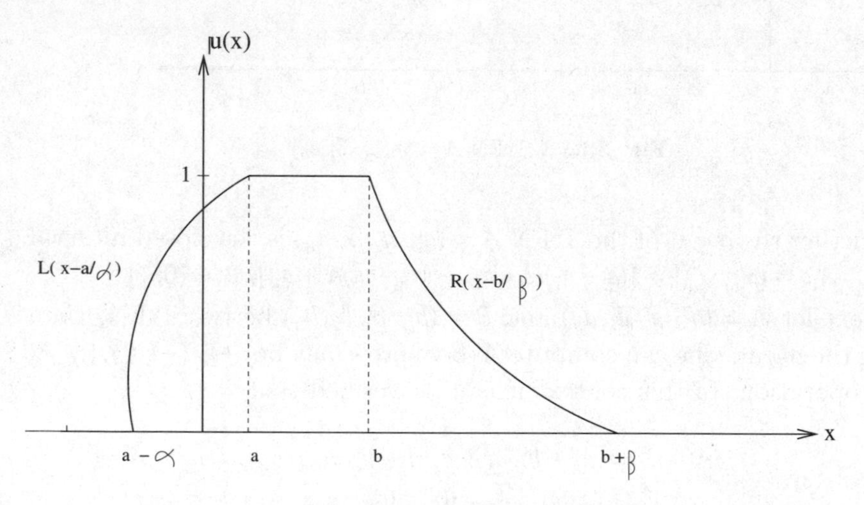

Fig. 3.7. L-R Fuzzy number *A*

Let $A = (a_1,\ b_1,\ \alpha,\ \beta)_{LR}$ and $B = (a_2,\ b_2,\ \gamma,\ \delta)_{LR}$ be two *L-R* fuzzy numbers. Then it can be verified that

$$A(+)B = (a_1 + a_2,\ b_1 + b_2,\ \alpha + \gamma,\ \beta + \delta)_{LR}$$

and

$$-A = -(a_1,\ b_1,\ \alpha,\ \beta) = (-b_1,\ -a_1,\ \beta,\ \alpha)_{RL}.$$

For defining the difference $A(-)B$ the original number B should be a R-L fuzzy number so that $-B$ becomes a L-R fuzzy number and one can compute $A(-)B$ as $A(+)(-B)$. Therefore for $A = (a_1,\ b_1,\ \alpha,\ \beta)_{LR}$ and $B = (a_2,\ b_2,\ \gamma,\ \delta)_{RL}$ we get

$$\begin{aligned} A(-)B &= (a_1,\ b_1,\ \alpha,\ \beta)_{LR}(+)(-(a_2,\ b_2,\ \gamma,\ \delta)_{RL}) \\ &= (a_1,\ b_1,\ \alpha,\ \beta)_{LR}(+)(-b_2,\ -a_2,\ \delta,\ \gamma)_{LR} \\ &= (a_1 - b_2,\ b_1 - a_2,\ \alpha + \delta,\ \beta + \gamma)_{LR}. \end{aligned}$$

Further, similar to TFNs and TrFNs, $A^{-1},\ A(\cdot)B,\ A(:)B$ etc. are not L-R fuzzy numbers in general and will need certain L-R approximations (Dubois and Prade [14] and [15]) if they are to be used as approximate L-R fuzzy numbers.

3.6 Ranking of fuzzy numbers

Ranking of fuzzy numbers is an important issue in the study of fuzzy set theory. Ranking procedures are also useful in various applications and one of them will be in the study of fuzzy mathematical programming and fuzzy games in later chapters. There are numerous methods proposed in the literature for the ranking of fuzzy numbers, some of them seem to be good in a particular context but not in general.

Our presentation here is going to be rather introductory as we describe only three simple methods for the ordering of fuzzy numbers. However this discussion will be continued in later chapters as well, in particular Chapter 10.

Ranking function (index) approach

Let $N(\mathbb{R})$ be the set of all fuzzy numbers in $\mathbb{R}$ and $A,\ B \in N(\mathbb{R})$. In this approach a suitable function $F : N(\mathbb{R}) \to \mathbb{R}$, called a *ranking function* or *ranking index* is defined and $F(A) \leq F(B)$ is treated as equivalent to $A(\leq)B$. Since $F(A) = F(B)$ will, in general, not mean $A = B$, this ranking (and many others to be studied in later chapters) is to be understood in the sense of equivalence classes only. Yager [87] proposed the following indices.

(i) $F_1(A) = \dfrac{\left(\int_{a_l}^{a_u} x\mu_A(x)dx\right)}{\left(\int_{a_l}^{a_u} \mu_A(x)dx\right)}$, where a_l and a_u are the lower and upper limits of the support of A. The value $F_1(A)$ represents the centroid of the fuzzy number $A \in N(\mathbb{R})$.
If, in particular, $A = (a_l,\ a,\ a_u)$ is a TFN then a_l and a_u are the lower and upper limits of the support of A and a is the modal value. In this case by actual substitution of the membership function of the TFN A, it can be verified that $F_1(A) = \dfrac{a_l + a + a_u}{3}$. Therefore given two TFNs $A = (a_l,\ a,\ a_u)$ and $B = (b_l,\ b,\ b_u)$, $A(\leq)B$ with respect to the index F_1 if and only if $(a_l + a + a_u) \leq (b_l + b + b_u)$.

(ii) $F_2(A) = \left(\int_0^{\alpha_{\max}} m[a_\alpha^L,\ a_\alpha^R]d\alpha\right)$, where α_{max} is the height of A, $A_\alpha = [a_\alpha^L,\ a_\alpha^R]$ is an α-cut, $\alpha \in (0,1]$, and $m[a_\alpha^L,\ a_\alpha^R]$ is the mean value of the elements of that α-cut. For a TFN $A = (a_l,\ a,\ a_u)$, $\alpha_{\max} = 1$ and $A_\alpha = [a_\alpha^L,\ a_\alpha^R] = [(a - a_l)\alpha + a_l,\ (a - a_u)\alpha + a_u]$.
Therefore
$$m[a_\alpha^L,\ a_\alpha^R] = \frac{(2a - a_l - a_u)\alpha + (a_l + a_u)}{2}$$
and
$$F_2(A) = \frac{(a_l + 2a + a_u)}{4}.$$
In view of the above we conclude that given two TFNs $A = (a_l,\ a,\ a_u)$ and $B = (b_l,\ b,\ b_u)$, $A(\leq)B$ with respect to the index F_2 if and only if $(a_l + 2a + a_u) \leq (b_l + 2b + b_u)$.

k-Preference index approach

This approach has been suggested by Adamo [1]. Let A be the given fuzzy number and $k \in [0,1]$. The *k-preference index* of A is defined as

$$F_k(A) = \max\{x : \mu_A(x) \geq k\}.$$

Now, using this k-preference index, for two fuzzy numbers $A,\ B \in N(\mathbb{R})$, $A(\leq)B$ with degree $k \in [0,1]$ if and only if $F_k(A) \leq F_k(B)$.

If $A = (a_l,\ a,\ a_u)$ and $B = (b_l,\ b,\ b_u)$ are two TFNs then for a given $k \in [0,1]$, $F_k(A) = ka + (1-k)a_u$ and $F_k(B) = kb + (1-k)b_u$. Therefore $A(\leq)B$ with respect to the k-preference index if and only if $\left(ka + (1-k)a_u\right) \leq \left(kb + (1-k)b_u\right)$.

Possibility theory approach

Dubois and Prade [16] studied the ranking of fuzzy numbers in the setting of possibility theory. To develop this, suppose we have two fuzzy number A and B. Then in accordance with the extension principle of Zadeh, the crisp inequality $x \leq y$ can be extended to obtain the truth value of the assertion that A is less than or equal to B, as follows:

$$T(A(\leq)B) = \sup_{x \leq y}\left(\min\left(\mu_A(x),\ \mu_B(y)\right)\right).$$

This truth value $T(A(\leq)B)$ is also called the *grade of possibility of dominance* of B on A and is denoted by $\mathtt{Poss}\ (A(\leq)B)$.

In a similar way, the grade (or degree) of possibility that the assertion "A is greater than or equal to B" is true, is given by

$$\mathtt{Poss}\ (A(\geq)B) = \sup_{x \geq y}\left(\min\left(\mu_A(x),\ \mu_B(y)\right)\right).$$

Also the degree of possibility that the assertion "A is equal to B" is denoted by $\mathtt{Poss}\ (A(=)B)$, and is defined as

$$\mathtt{Poss}\ (A(=)B) = \sup_{x}\left(\min\left(\mu_A(x),\ \mu_B(x)\right)\right).$$

The above discussion motivates us to define $A(\leq)B$ if and only if $\mathtt{Poss}\ (A(\leq)B) \geq \mathtt{Poss}\ (B(\leq)A)$. Here it may be noted that for the case when $A = (a_l,\ a,\ a_u)$ and $B = (b_l,\ b,\ b_u)$ are TFN then $a \leq b$ gives $\mathtt{Poss}\ (A(\leq)B) = 1$ and $\mathtt{Poss}\ (B(\leq)A) =$ height $(A \cap B) \leq 1$.

Therefore for the case of TFNs it can be defined that $A(\leq)B$ with respect to $\mathtt{Poss}\ (A(\leq)B)$ approach if $a \leq b$.

Related with the number "$\mathtt{Poss}\ (A(\leq)B)$" there is another number "$\mathtt{Necc}(A(\leq)B)$" which measures the grade (or degree) of necessity of dominance of B on A, given by

$$\mathtt{Necc}(A(\leq)B) = 1 - \mathtt{Poss}\ (A(\geq)B).$$

The number "$\mathtt{Necc}(A(\leq)B)$" can also be used for ranking of fuzzy numbers. For this, we can define $A(\leq)B$ if and only if $\mathtt{Necc}(A(\leq)B) \geq \mathtt{Necc}(B(\leq)A)$.

In case $A = (a_l, a, a_u)$ and $B = (b_l, b, b_u)$ are TFN then by actual computation of $\mathtt{Necc}(A(\geq)B)$ it can be defined that $A(\leq)B$ with respect to $\mathtt{Necc}(A(\leq)B)$ approach if $a + a_l \leq b + b_l$.

The "Poss" and "Necc" as defined above are special type of *fuzzy measures* which are discussed in somewhat greater detail in Chapter 10.

3.7 Conclusions

In this chapter we have presented a very basic but brief discussion on fuzzy numbers and fuzzy arithmetic. Since ranking of fuzzy numbers is an important aspect in the study of fuzzy mathematical programming and fuzzy games, we shall continue our discussion on this topic in later chapters as well.

4

Linear and quadratic programming under fuzzy environment

4.1 Introduction

For the linear programming problems (LPPs) in the crisp scenario, the aim is to maximize or minimize a linear objective function under linear constraints. But in many practical situations, the decision maker may not be in a position to specify the objective and/or constraint functions precisely but rather can specify them in a "fuzzy sense". In such situations, it is desirable to use some fuzzy linear programming type of modeling so as to provide more flexibility to the decision maker. Since the fuzziness may appear in a linear programming problem in many ways (e.g. the inequalities may be fuzzy, the goals may be fuzzy or the problem parameters c, A, b may be in terms of fuzzy numbers), the definition of fuzzy linear programming problem is not unique. This chapter aims to study various models of fuzzy linear programming problems and their possible extensions for fuzzy quadratic programming.

This chapter consists of seven main sections, namely, *decision making under fuzzy environment and fuzzy linear programming, linear programming problems with fuzzy inequalities and crisp objective function, linear programming problems with crisp inequalities and fuzzy objective function, linear programming problems with fuzzy inequalities and fuzzy objective function, quadratic programming under fuzzy environment: symmetric and non symmetric models, a two phase approach for solving fuzzy linear programming problems, and linear goal programming under fuzzy environment.*

4.2 Decision making under fuzzy environment and fuzzy linear programming

A fuzzy decision making model is characterized by a set X of possible actions/alternatives, and a set of goals G_i $(i = 1, 2, \dots, p)$, along with a set of constraints C_j $(j = 1, 2, \dots, n)$, each of which is expressed by a fuzzy set on X. For such a model of decision making, Bellman and Zadeh [9] in their pioneering work, proposed that a fuzzy decision is determined by an appropriate aggregation of the fuzzy sets G_i $(i = 1, 2, \dots, m)$ and C_j $(j = 1, 2, \dots, n)$. In this approach the symmetry between goals and constraints is the main feature. Keeping this in mind, they (Bellman and Zadeh [9]) suggested the aggregation operator to be the fuzzy intersection. Thus a *fuzzy decision* D could be defined as the fuzzy set $D = (G_1 \cap G_2 \cap \dots \cap G_p) \cap (C_1 \cap C_2 \cap \dots \cap C_n)$, i.e $\mu_D : X \to [0, 1]$ given by $\mu_D(x) = \min\limits_{i,j} \left(\mu_{G_i}(x), \mu_{C_j}(x)\right)$.

Once the fuzzy decision D is known, we can define $x^* \in X$ to be an *optimal decision* if $\mu_D(x^*) = \max\limits_{x} \mu_D(x)$. Another possibility could be to choose an α $(0 < \alpha < 1)$ and determine all points $x^* \in X$ for which $\mu_D(x^*) \geq \alpha$. These decisions x^* will have at least α degree of membership value.

Example 4.2.1. (Zimmermann [91]). As an illustration let us consider the fuzzy decision problem in which we have to find a real number x which is in the vicinity of 15 (fuzzy constraint C) and is *substantially larger than* 10 (fuzzy goal G). For this problem, the goal and the constraint can be expressed in terms of their membership functions as follows

$$\mu_G(x) = \begin{cases} 0 & , x \leq 10, \\ \left(1 + (x - 10)^{-2}\right)^{-1} & , x > 10, \end{cases}$$

and,

$$\mu_C(x) = \left(1 + (x - 15)^4\right)^{-1}.$$

Here it may be noted that since in this approach, goals and constraints are symmetric, we may denote $\mu_G(x)$ by $\mu_C(x)$ and vice versa. Now as per the Bellman and Zadeh [9] approach, the fuzzy decision D equals $G \cap C$, i.e.

$$\mu_D(x) = \min\left(\mu_G(x), \mu_C(x)\right).$$

The following diagram depicts this fuzzy decision making problem and identifies the optimal solution x^*.

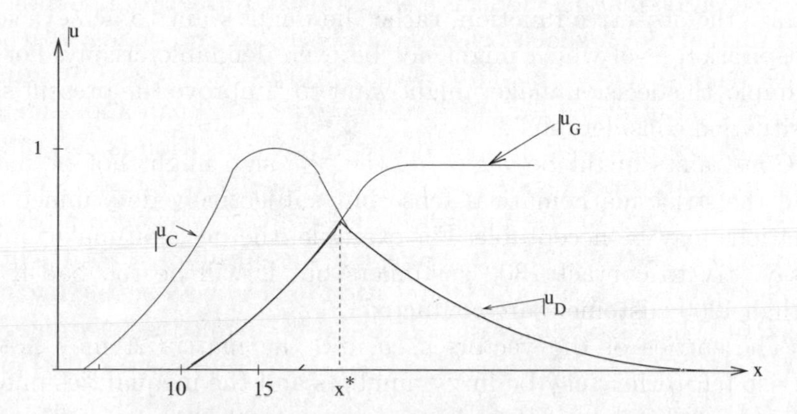

Fig. 4.1. graphical representation of μ_D and x^*

We now try to understand and conceptualize *fuzzy linear programming models* in the light of Bellman and Zadeh principle as discussed above. The classical linear programming problem aims to find the minimum or maximum of a linear function under constraints which are represented by linear inequalities or equations. The most typical linear programming problem is stated as

$$(LP) \qquad \max \quad c^T x$$
$$\text{subject to,}$$
$$Ax \leq b,$$
$$x \geq 0,$$

where $x \in \mathbb{R}^n$, $c \in \mathbb{R}^n$, $b \in \mathbb{R}^m$, and $A \in \mathbb{R}^m \times \mathbb{R}^n$.

In the *decision making terminology*, x is referred as a vector of *decision variables*, b as a vector of *available resources*, c as a vector of *cost coefficients* and A as the *constraint matrix*.

In the above formulation, it is assumed that all entries of A, b and c are crisp numbers, "$\leq$" is defined in the crisp sense and "max" is a strict imperative. However, in many practical situations it may not be reasonable to require that the constraints or the objective function in linear programming problem be specified in precise crisp terms. In such situations, it is desirable to use some type of fuzzy modeling and this leads to the concept of fuzzy linear programming.

When decision is to be made in a fuzzy environment, many possible modifications of the above linear programming model exist, such as,

(i) The decision maker might not really want to maximize or minimize the objective function, rather he might want to achieve some aspiration level which might not be even definable crisply. For example, the decision maker might want to "improve the present sales situation considerably".
(ii) Constraints might be vague i.e. the "$\leq$" sign might not be meant in the strict mathematical sense but subjectively determined violations may be acceptable. For example, the decision maker might say "try to contact 1300 customers but it will be too bad if less than 1200 customers are contacted".
(iii) The entries of the vectors c, b and the matrix A may not be crisp but rather may be fuzzy numbers and the inequalities may be interpreted in terms of ranking of fuzzy numbers.

Thus in view of the above, fuzzy linear programming models are not uniquely defined as it will very much depend upon the type of fuzziness and its specification as prescribed by the decision maker. Therefore the class of fuzzy linear programming problems can be broadly classified as

(i) linear programming problem with fuzzy inequalities and crisp objective function,
(ii) linear programming problems with crisp inequalities and fuzzy objective function,
(iii) linear programming problems with fuzzy inequalities and fuzzy objective function, and
(iv) linear programming problems with fuzzy resources and fuzzy coefficient, also termed as linear programming problems with fuzzy parameters, i.e. elements of c, b and A are fuzzy numbers.

The class of fuzzy linear programming problems can also be classified as *symmetric or non symmetric*. The *symmetric* models are based on the definition of fuzzy decision as proposed by Bellman and Zadeh [9] and discussed here in Section 4.2. The basic feature being the symmetry of objectives and constraints, this approach gives the decision set as a fuzzy set resulting from the intersection of the fuzzy sets corresponding to the objective and constraints. On the other hand, the *non symmetric* models keep distinction between the objective and constraints. Here usually two approaches are followed. In the first approach a fuzzy set of *decisions* is determined and then the (crisp) objective function is "maximized" over this fuzzy set. This approach leads to a parametric linear programming problem. In the second approach, after determining the fuzzy set of decisions, a suitable membership function of the objective function is determined and then the problem is solved similar to the symmetric case.

In the next two sections now, we discuss some of the most common models of fuzzy linear programming problems. Though these models look simple, they have wide applications and have been used extensively in the literature. The class of fuzzy linear programming problems for which entries of c, b and A are fuzzy numbers, are not discussed here, instead they are studied in Chapters 6, 7, 8 and 10.

4.3 LPPs with fuzzy inequalities and crisp objective function

The general model of a linear programming problem with fuzzy inequalities and crisp objective function is as follows:

$$\begin{aligned} &\max \quad c^T x \\ &\text{subject to,} \\ &\qquad A_i x \lesssim b_i, \; (i = 1, 2 \ldots, m), \\ &\qquad x \geq 0, \end{aligned}$$

where $\lesssim$ is called "fuzzy less than or equal to" and is to be understood in terms of a suitably chosen membership function.

In Verdegay's terminology, such fuzzy linear programming problems are said to be of "type $P1$" and therefore we shall now onwards denote the above problems as ($P1$-FLP).

Verdegay's approach : a non symmetric model

Verdegay ([74], [75]) showed that the problem ($P1$-FLP) is equivalent to a crisp parametric linear programming problem and therefore we can use parametric programming methods to solve such fuzzy linear programming problems. Here the fuzzy constraints are transformed into crisp constraints by choosing appropriate membership function for each constraint. To motivate for a meaningful choice of membership function, it is argued that if $A_i x \leq b_i$ then the ith constraint is absolutely satisfied, where as if $A_i x \geq b_i + p_i$, where p_i is the maximum tolerance from b_i, as determined by the decision maker, then the ith constraint is absolutely violated. For $A_i x \in (b_i,\ b_i + p_i)$, the membership function is monotonically decreasing. If this decrease is along a linear function then it makes sense to choose the membership function of the ith constraint ($i = 1, 2, \ldots, m$) as

$$\mu_i(A_i x) = \begin{cases} 1 & , A_i x < b_i, \\ 1 - \dfrac{A_i x - b_i}{p_i} & , b_i \leq A_i x \leq b_i + p_i, \\ 0 & , A_i x > b_i + p_i, \end{cases}$$

where A_i ($i = 1, 2, \ldots, m$) denotes the ith row of A.

Now, for $\alpha \in [0, 1]$ let $X_\alpha = \{\ x \in \mathbb{R}^n :\ x \geq 0 \text{ and } \mu_i(A_i x) \geq \alpha,\ (i = 1, 2, \ldots, m)\ \}$ then the problem (*P*1-*FLP*) is equivalent to

$$\begin{aligned} &\max \quad c^T x \\ &\text{subject to,} \\ &\qquad x \in X_\alpha. \end{aligned}$$

We can substitute the expression for the membership functions $\mu_i(A_i x)$ and obtain the following problem:

$$\begin{aligned} (LP)_\alpha \qquad &\max \quad c^T x \\ &\text{subject to,} \\ &\quad A_i x \leq b_i + (1 - \alpha) p_i,\ (i = 1, 2, \ldots, m), \\ &\quad x \geq 0, \\ &\quad \alpha \in [0, 1], \end{aligned}$$

which is equivalent to a standard parametric linear programming problem, with $\theta = (1 - \alpha)$. Thus the fuzzy linear programming problem (*P*1-*FLP*) can be solved by solving an equivalent crisp parametric linear programming problem. Here, it may be noted that we have an optimal solution for each $\alpha \in [0, 1]$, so the solution with α grade of membership is actually fuzzy.

Werners' approach : a symmetric model

Werners [79] proposed that for the problems of the type (*P*1-*FLP*), the objective function should be fuzzy because of fuzzy inequality constraints. Further, to construct a membership function for the objective function, he suggested to solve the following two linear programming problems ($LP(b)$) and ($LP(b + p)$)

$$(LP(b)) \qquad \max \quad c^T x \quad \text{subject to,} \quad Ax \le b, \; x \ge 0,$$

$$(LP(b+p)) \qquad \max \quad c^T x \quad \text{subject to,} \quad Ax \le b + p, \; x \ge 0.$$

Here as before, $p = (p_1, p_2, \ldots, p_m)^T$ is the vector of tolerances for the m constraints of (*P1-FLP*). Let Z_0 and Z_1 be optimal values of $(LP(b))$ and $(LP(b+p))$ respectively.

We can now construct a continuously nondecreasing linear membership function μ_0 for the objective function by using Z_0 and Z_1 as follows

$$\mu_0(c^T x) = \begin{cases} 1 & , c^T x > Z_1, \\ 1 - \dfrac{Z_1 - c^T x}{Z_1 - Z_0} & , Z_0 \le c^T x \le Z_1, \\ 0 & , c^T x > Z_0. \end{cases}$$

The membership functions of the constraints are the same as in Verdegay's approach, i.e.

$$\mu_i(A_i x) = \begin{cases} 1 & , A_i x < b_i, \\ 1 - \dfrac{A_i x - b_i}{p_i} & , b_i \le A_i x \le b_i + p_i, \\ 0 & , A_i x > b_i + p_i. \end{cases}$$

Now using the above membership functions, μ_i $(i = 0, 1, \ldots, m)$ and following Bellman and Zadeh principle, the problem (*P1-FLP*) is solved by solving the following crisp linear programming problem

$$\begin{aligned} \max \quad & \alpha \\ \text{subject to,} \quad & \\ & \mu_0(x) \ge \alpha, \\ & \mu_i(x) \ge \alpha, \quad (i = 1, 2, \ldots, m), \\ & \alpha \in [0, 1], \\ & x \ge 0, \end{aligned}$$

which on substitution for μ_i $(i = 0, 1, 2, \ldots, m)$ becomes

$$\begin{array}{ll} \max & \alpha \\ \text{subject to,} & \\ & c^T x \geq Z_1 - (1 - \alpha)(Z_1 - Z_0), \\ & A_i x \leq b_i + (1 - \alpha) p_i, \\ & \alpha \in [0, 1], \\ & x \geq 0. \end{array}$$

4.4 LPPs with crisp inequalities and fuzzy objective functions

A linear programming problem with crisp inequalities and fuzzy objective function is described as

$$\begin{array}{ll} \tilde{\max} & c^T x \\ \text{subject to,} & \\ & Ax \leq b, \\ & x \geq 0, \end{array}$$

and then following Verdegay's notations ([74], [75]), it is denoted by (*P*2-*FLP*).

Let the membership function of the fuzzy objective be given by

$$\phi(c) = \inf_j \phi_j(c_j),$$

where $\phi_j : \mathbb{R} \to [0, 1]$, $j = 1, 2, \ldots, n$.

Verdegay [74] argued that a fuzzy solution of (*P*2-*FLP*) could be found by solving the crisp linear programming problem

$$(LP)_{(\phi,\,\alpha)} \qquad \begin{array}{ll} \max & c^T x \\ \text{subject to,} & \\ & \phi(c) \geq (1 - \alpha), \\ & Ax \leq b, \\ & x \geq 0, \\ & \alpha \in [0, 1]. \end{array}$$

In general, it may not be easy to solve $(LP)_{(\phi,\,\alpha)}$ so we express it an equivalent way as per the theorem given below. For this we introduce the problem $(LP)_\beta$ as follows

$(LP)_\beta$ max $\eta(\beta)^T x$
subject to,

$$\begin{aligned} Ax &\le b, \\ x &\ge 0, \\ \beta &\in [0,1], \end{aligned}$$

where $\eta(\beta)$ is a vector function $\left(\eta_1(\beta), \ldots, \eta_j(\beta)\right)$ with $\eta_j : [0,1] \to \mathbb{R}$.

Theorem 4.4.1 *In the problem (P2-FLP), let the membership functions $\phi_j : \mathbb{R} \to [0,1]$, $(j = 1, 2, \ldots, n)$, be continuous and strictly monotone. Then the fuzzy solution of (P2-FLP) is given by the parametric solution of the parametric linear programming problem $(LP)_\beta$.*

Proof. We have to solve the following problem to obtain the solution of *(P2-FLP)*
$(LP)_{(\phi,\,\alpha)}$ max $c^T x$
subject to,

$$\begin{aligned} \phi(c) &\ge 1 - \alpha, \\ Ax &\le b, \\ x &\ge 0, \\ \alpha &\in [0,1], \end{aligned}$$

where $\phi(c) = \inf_j \left(\phi_j(c_j)\right)$. But ϕ_j is continuous and strictly monotone, giving that ϕ_j^{-1} exists, and $\phi_j(c_j) \ge (1-\alpha) \Rightarrow c_j \ge \phi_j^{-1}(1-\alpha)$. Therefore the problem $(LP)_{(\phi,\,\alpha)}$ can be written as

$$\max \quad \sum_{j=1}^{n} c_j x_j$$

subject to,

$$\begin{aligned} c_j &\ge \phi_j^{-1}(1-\alpha), \ (j = 1, 2, \ldots, n), \\ Ax &\le b, \\ x &\ge 0, \\ \alpha &\in [0,1], \end{aligned}$$

which is equivalent to

$$\max \quad \sum_{j=1}^{n} c_j x_j$$

subject to,

$$\begin{aligned} c_j &= \phi_j^{-1}(1-\alpha), (j = 1, 2, \ldots, n), \\ Ax &\leq b, \\ x &\geq 0, \\ \alpha &\in [0, 1]. \end{aligned}$$

Thus the fuzzy linear programming (*P2-FLP*) can be solved by solving the (crisp) linear programming problem

$$\begin{aligned} \max \quad & \sum_{j=1}^{n} \left(\phi_j^{-1}(1-\alpha)\right) x_j \\ \text{subject to,} \quad & \\ & Ax \leq b, \\ & x \geq 0, \\ & \alpha \in [0, 1], \end{aligned}$$

which is the same as the problem $(LP)_\beta$ with $\beta = (1-\alpha)$ and $\eta_j(\beta) = \phi_j^{-1}(1-\alpha)$.

Remark 4.4.1. Since the parameterizations used here do not effect any change in the feasible region, the problems of the type (*P2-FLP*) are probably easier to solve than the problems of type (*P1-FLP*). In fact there is a complete duality between these two types of linear programming problems as any problem of type (*P1-FLP*) could be transformed into a problem of type (*P2-FLP*) and vice versa. This suggests a dual approach of solving fuzzy linear programming problems which we shall discuss in Chapter 5.

Example 4.4.2. (Verdegay[74]). Consider the fuzzy linear programming problem (*P2-FLP*) as

$$\begin{aligned} \tilde{\max} \quad & c_1 x_1 + c_2 x_2 \\ \text{subject to,} \quad & \\ & 3x_1 - x_2 \leq 2, \\ & x_1 + 2x_2 \leq 3, \\ & x_1, x_2 \geq 0, \end{aligned}$$

where $c_2 = 75$ (crisp) and the membership function ϕ_1 for c_1 is given by

$$\phi_1(c_1) = \begin{cases} 0 & , c_1 < 40, \\ \dfrac{(c_1 - 40)^2}{5625} & , 40 \leq c_1 \leq 115, \\ 1 & , c_1 > 115. \end{cases}$$

To solve the given fuzzy linear programming problem, we have to solve the crisp problem

$$\begin{aligned} \max \quad & \sum_{j=1}^{n} \left(\phi_j^{-1}(1-\alpha)\right)x_j \\ \text{subject to,} \quad & \\ & Ax \le b, \\ & x \ge 0. \end{aligned}$$

Since here, only one membership function, namely $\phi_1(c)$, is taking part, we have to solve the problem

$$\begin{aligned} \max \quad & \left(\phi_1^{-1}(1-\alpha)\right)x_1 + 75x_2 \\ \text{subject to,} \quad & \\ & 3x_1 - x_2 \le 2, \\ & x_1 + 2x_2 \le 3, \\ & x_1, x_2 \ge 0. \end{aligned}$$

But from the definition of $\phi_1(c)$, we have $\phi_1^{-1}(1-\alpha) = \left(40 + 75\sqrt{(1-\alpha)}\right)$ and therefore the above problem becomes

$$\begin{aligned} \max \quad & \left(40 + 75\sqrt{(1-\alpha)}\right)x_1 + 75x_2 \\ \text{subject to,} \quad & \\ & 3x_1 - x_2 \le 2, \\ & x_1 + 2x_2 \le 3, \\ & x_1, x_2 \ge 0, \\ & \alpha \in [0,1]. \end{aligned}$$

The optimal solution of the above parametric linear programming problem is $x_1^* = 1$, $x_2^* = 1$ and therefore the fuzzy solution, i.e. the fuzzy set of the values of the objective function, is obtained as

$$\left\{(75 + c_1), \frac{(c_1 - 40)^2}{5625}\right\}, \ 40 \le c_1 \le 115.$$

4.5 LPPs with fuzzy inequalities and objective function

The general model of a linear programming problem with fuzzy objective and fuzzy constraints is formulated as:

$$\begin{aligned} \tilde{\max} \quad & c^T x \\ \text{subject to,} \quad & \end{aligned}$$

$$Ax \lesssim b,$$
$$x \geq 0,$$

where we need to explain the fuzzifiers "$\lesssim$" and $\tilde{\max}$ in the context of the model under consideration. In the terminology of Verdegay ([74], [75]), we shall denote the above fuzzy linear programming problem as (*P3-FLP*).

Zimmermann's approach : a symmetric model

In this approach, the fuzzy constraints are handelled exactly in the same manner as in the earlier case described in Section 4.3. But for the objective function the fuzzifier $\tilde{\max}$ is understood in the sense of the satisfaction of an aspiration level Z_0 as best as possible.

In view of the above, we can describe the problem (*P3-FLP*) as follows

(*FSI*) Find x such that

$$c^T x \gtrsim Z_0,$$
$$Ax \lesssim b,$$
$$x \geq 0.$$

To solve the above problem, we should first choose an appropriate membership function for each of the fuzzy inequality and then employ Bellman and Zadeh principle to identify the fuzzy decision. In particular, let μ_0 denote the membership function for the objective function and μ_i $(i = 1, 2, \ldots, m)$ denote the membership function for the ith constraint. Let p_0 and p_i $(i = 1, 2, \ldots, m)$ respectively be the permissible tolerances for the objective function and the ith constraint. Then we may decide μ_0 and μ_i $(i = 1, 2, \ldots, m)$ to be nondecreasing and continuous linear membership functions as per the choice given below

$$\mu_0(c^T x) = \begin{cases} 1 & , c^T x > Z_0, \\ 1 - \dfrac{Z_0 - c^T x}{p_0} & , Z_0 - p_0 \leq c^T x \leq Z_0, \\ 0 & , c^T x < Z_0 - p_0, \end{cases}$$

and

$$\mu_i(A_i x) = \begin{cases} 1 & , A_i x < b_i, \\ 1 - \dfrac{A_i x - b_i}{p_i} & , b_i \leq A_i x \leq b_i + p_i, \\ 0 & , A_i x > b_i + p_i. \end{cases}$$

Now we use the Bellman and Zadeh's principle to identify the fuzzy decision to solve the fuzzy system of inequalities (*FSI*) corresponding to the problem (*P3-FLP*). This leads to the following crisp linear programming problem,

$$
\begin{aligned}
&\max \quad \alpha\\
&\text{subject to,}\\
&\mu_0(c^T x) = \left(1 - \frac{Z_0 - c^T x}{p_0}\right) \geq \alpha,\\
&\mu_i(A_i x) = \left(1 - \frac{A_i x - b_i}{p_i}\right) \geq \alpha, \quad (i = 1, 2, \ldots, m),\\
&\alpha \in [0, 1],\\
&x \geq 0,
\end{aligned}
$$

or

$$
\begin{aligned}
&\max \quad \alpha\\
&\text{subject to,}\\
&c^T x \geq Z_0 - (1 - \alpha) p_0,\\
&A_i x \leq b_i + (1 - \alpha) p_i, \quad (i = 1, 2, \ldots, m),\\
&\alpha \in [0, 1],\\
&x \geq 0,
\end{aligned}
$$

If (x^*, λ^*) is an optimal solution of the above (crisp) linear programming problem then x^* is said to be an optimal solution of the problem (*P3-FLP*) and λ^* is the degree up to which the aspiration level Z_0 of the decision maker is met.

Remark 4.5.1. If the problem (*P3-FLP*) has fuzzy as well as crisp constraints, then in the equivalent (crisp) linear programming problem, the original crisp constraints will not have any change as for them the tolerances are zero.

Example 4.5.2. (Zimmermann [90]). Consider the fuzzy linear programming problem (*P3-FLP*) as

$$
\begin{aligned}
&\widetilde{\max} \quad x_1 + x_2\\
&\text{subject to,}\\
&-x_1 + 3x_2 \lesssim 21,\\
&x_1 + 3x_2 \lesssim 27,\\
&4x_1 + 3x_2 \lesssim 45,\\
&3x_1 + x_2 \leq 30,\\
&x_1, x_2 \geq 0,
\end{aligned}
$$

Here it is given that $Z_0 = 14.5$, $p_0 = 2$, $p_1 = 3$, $p_2 = 6$, and $p_3 = 6$. Following Zimmermann's approach, to solve the given problem (*P3-FLP*), we need to solve the following (crisp) linear programming problem:

$$\begin{aligned} \max \quad & \alpha \\ \text{subject to,} \quad & \\ & x_1 + x_2 \geq 14.5 - 2(1-\alpha), \\ & -x_1 + 3x_2 \leq 21 + 3(1-\alpha), \\ & x_1 + 3x_2 \leq 27 + 6(1-\alpha), \\ & 4x_1 + 3x_2 \leq 45 + 6(1-\alpha), \\ & 3x_1 + x_2 \leq 30, \\ & \alpha \leq 1, \\ & x_1, x_2, \alpha \geq 0. \end{aligned}$$

i.e

$$\begin{aligned} \max \quad & \alpha \\ \text{subject to,} \quad & \\ & 2\alpha - x_1 - x_2 \leq -12.5, \\ & 3\alpha - x_1 + 3x_2 \leq 24, \\ & 6\alpha + x_1 - 3x_2 \leq 33, \\ & 6\alpha + 4x_1 + 3x_2 \leq 51, \\ & \alpha \leq 1, \\ & x_1, x_2, \alpha \geq 0. \end{aligned}$$

By using Simplex algorithm, we can solve the above linear programming problem and obtain the solution $x_1^* = 6$, $x_2^* = 7.75$, $z^* = 13.75$ and $\lambda^* = 0.625$.

Chanas' approach : a non symmetric model

We have seen that for solving fuzzy linear programming problems of type (*P3-FLP*) by Zimmermann's approach, we need the aspiration level Z_0 for the objective function and its associated tolerance p_0. Chanas [12] argued that because of lack of knowledge about the fuzzy feasible region, it may not be easy to specify the aspiration level Z_0, and its tolerance p_0. He therefore suggested first to solve the following problem (*P1-FLP*), and then present the results to the decision maker to determine Z_0 and p_0.

$$(P1-FLP) \qquad \max \quad z = c^T x$$
$$\text{subject to,}$$

$$Ax \lesssim b,$$
$$x \geq 0,$$

For the given tolerances p_i $(i = 1, 2, \ldots, m)$, we may choose the membership functions μ_i as before, i.e.

$$\mu_i(A_i x) = \begin{cases} 1 & , A_i x < b_i, \\ 1 - \dfrac{A_i x - b_i}{p_i} & , b_i \leq A_i x \leq b_i + p_i, \\ 0 & , A_i x > b_i + p_i. \end{cases}$$

Employing the Verdegay's approach, the problem (*P*1-*FLP*) is equivalent to the following parametric linear programming problem $(LP)_\theta$:

$$(LP)_\theta \qquad \max \quad z = c^T x$$
$$\text{subject to,}$$
$$Ax \leq b + \theta p,$$
$$x \geq 0,$$
$$\theta \in [0, 1],$$

where $\theta = (1 - \alpha)$ is a parameter and $p = (p_1, p_2, \ldots, p_m)^T$ is the vector of tolerances for each of the m fuzzy constraints.

Let for a given θ, $x^*(\theta)$ be an optimal solution of $(LP)_\theta$ and the corresponding optimal value be $z^*(\theta)$. Then the constraint $Ax \leq b + \theta p$ means, $A_i x \leq b_i + (1 - \alpha)p_i$, $(i = 1, 2, \ldots, m)$, which is equivalent to the statement that for each $i = 1, 2, \ldots, m$, $\mu_i\big(A_i x^*(\theta)\big) \geq \alpha = (1 - \theta)$, holds. Also there exists at least one i for which $\mu_i\big(A_i x^*(\theta)\big) = \alpha = (1 - \theta)$. This is because in $(LP)_\theta$ at least one constraint should be active at $x^*(\theta)$ as $x^*(\theta)$ can not be an interior point of the feasible region.

Therefore the common degree of satisfaction of these constraints is the minimum of $\mu_i\big(A_i x^*(\theta)\big)$ over i, i.e. $\mu_c\big(Ax^*(\theta)\big) = \min_i \mu_i\big(A_i x^*(\theta)\big) = (1 - \theta)$.

Hence, for every θ we obtain an optimal solution $x^*(\theta)$ with the optimal value $z^*(\theta)$, (if one exists) which satisfies jointly the constraints with degree $(1 - \theta)$. This optimal solution of $(LP)_\theta$ is then presented to the decision maker who, in the light of this information, choose Z_0 and corresponding p_0. We can then construct the membership function μ_0 of the objective function as follows

$$\mu_0(c^T x^*(\theta)) = \begin{cases} 1 & , c^T x^*(\theta) > Z_0, \\ 1 - \dfrac{Z_0 - c^T x^*(\theta)}{p_0} & , Z_0 - p_0 \leq c^T x^*(\theta) \leq Z_0, \\ 0 & , c^T x^*(\theta) < Z_0 - p_0. \end{cases}$$

Now from the expression of $\mu_0\big(c^T x^*(\theta)\big)$, we observe that it is a piecewise linear function of θ as depicted in the Figure 4.2. Also the common degree of satisfaction of the constraints is $\mu_c\big(Ax^*(\theta)\big) = (1-\theta)$, which is also depicted in Figure 4.2. Therefore the optimal solution of the problem (*P3-FLP*) is taken $x^*(\theta^*)$ with the optimal value $z^*(\theta^*)$ where θ^* is chosen such that $\mu_D(\theta^*) = \max_{\theta} \mu_D(\theta) = \max_{\theta} \Big(\min\big(\mu_0(\theta),\, \mu_c(\theta)\big)\Big)$.

Here $\mu_0(\theta)$ and $\mu_c(\theta)$ are in fact $\mu_0\big(x^*(\theta)\big)$ and $\mu_c\big(x^*(\theta)\big)$ respectively. The optimal choice of θ^* and associated value $\mu_D(\theta^*)$ are again shown in Figure 4.2.

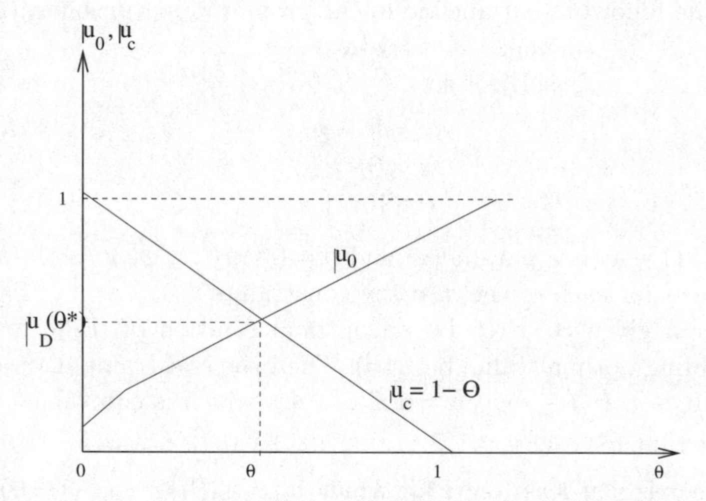

Fig. 4.2. The membership functions μ_0, μ_c and θ^*.

4.6 Quadratic programming under fuzzy environment

In contrast with the vast literature on modeling and solution procedures for linear programming in a fuzzy environment, the studies in general mathematical programming (even in quadratic programming) under fuzzy environment and its solution are pretty rare. In this section we make an attempt to study quadratic programming under fuzzy environment along the lines of fuzzy linear programming as discussed in Sections 4.3, 4.4, and 4.5. The results presented here are based on [5].

In the literature, a classical quadratic programming problem is stated as follows:

$$(QPP) \quad \min \quad c^T x + \frac{1}{2} x^T H x$$
subject to,
$$Ax \leq b,$$
$$x \geq 0,$$

where $c \in \mathbb{R}^n$, $x \in \mathbb{R}^n$, $A \in \mathbb{R}^m \times \mathbb{R}^n$ is an $m \times n$ matrix, and $b \in \mathbb{R}^m$. Also, H is a symmetric $n \times n$ matrix which is positive semi definite. Thus, the quadratic form $x^T H x$ is convex and hence the objective function is convex. We also make the assumption that the feasible region of (QPP) is bounded.

Several methods, e.g. Wolfe [80] and Beale [3], are available in the literature for solving such quadratic programming problems. However to study quadratic programming in a fuzzy environment from a solution point of view we require a solution procedure for solving a linear programming problem with one quadratic constraint, i.e. a special optimization problem in which the objective function and all constraints are linear except one constraint which is quadratic. Van de Panne [73] suggested a finite step method that uses linear programming and parametric quadratic programming to solve such problems. We now describe Van de Panne's method very briefly.

The mathematical model of such a special optimization problem is as follows:

$$(SLP) \quad \max \quad c^T x$$
subject to,
$$d^T x + \frac{1}{2} x^T Q x \leq \beta,$$
$$Ax \leq b,$$
$$x \geq 0,$$

where $d \in \mathbb{R}^n$, $Q \in \mathbb{R}^n \times \mathbb{R}^n$ a positive semi definite matrix, and $\beta \in \mathbb{R}$ is known in advance. Other notations are same as used in (QPP).

Van de Panne [73] developed the following *two phase method* to solve the optimization problem (SLP) in a finite number of steps.

Phase 1: In (SLP), we ignore the quadratic constraint to get the following ordinary linear programming problem, which is then solved under the assumption that it has an optimal solution.

$$(LP) \quad \max \quad c^T x$$
subject to,
$$Ax \leq b,$$
$$x \geq 0.$$

Let x_0 be optimal solution of the above linear programming problem (LP). If this optimal solution satisfies the quadratic constraint, that is if

$$d^T x_0 + \frac{1}{2} x_0^T Q x_0 \leq \beta,$$

then we have obviously found the optimal solution of the original problem, i.e. $x = x_0$ is the optimal solution to (SLP).

If however x_0 does not satisfy the quadratic constraint, that is if

$$d^T x_0 + \frac{1}{2} x_0^T Q x_0 > \beta,$$

then we move on to next phase of the technique.

Phase 2: In this phase we construct the following problem $(QPP)_\lambda$

$$(QPP)_\lambda \qquad \max \quad d^T x + \frac{1}{2} x^T Q x$$

subject to,

$$Ax \leq b,$$
$$c^T x \geq \lambda,$$
$$x \geq 0,$$

where λ is a parameter and it is assigned different values in the course of computations assuming that $\lambda_0 = c^T x_0$ to start with.

Van de Panne [73] solved the problem $(QPP)_\lambda$ by decreasing λ parametrically from λ_0 to lower values. Then Phase 2 can terminate in one of the two ways:

(*i*) it may terminate when for a certain value λ^* of λ, corresponding to $x = x^*$, the objective function has become equal to β. In this case an optimal solution to (SLP) has been found.

(*ii*) it may terminate when for a certain value of λ, say λ^{**}, the constraint $c^T x \geq \lambda$ ceases to be binding for the optimal solution of the problem $(QPP)_\lambda$ with $d^T x + \frac{1}{2} x^T Q x$ being still larger than β. This means that for no value of λ, a solution exists giving a minimum value of the objective $d^T x + \frac{1}{2} x^T Q x$ less than or equal to β.

In this case, in (SLP), the quadratic constraint is incompatible with the linear constraints and no feasible solution to (SLP) exists.

Now when we fuzzify the classical quadratic programming problem (QPP) as in the case of linear programming , then we get the two models, namely, the *symmetric fuzzy* and *non symmetric fuzzy quadratic programming problems.* In order to find a solution we transform each of them into the crisp programming problem of the type (SLP).

The fuzzy version of the classical quadratic program (QPP) can be written as

$$(FQP) \qquad \widetilde{\min} \qquad Z = c^T x + \frac{1}{2} x^T H x$$

subject to,

$$A_i x \lesssim b_i, \ (i = 1, 2, \ldots, m),$$
$$x \geq 0,$$

where as before, the fuzzification is to be understood in terms of the appropriate membership functions.

Symmetric fuzzy quadratic programming

We consider the following symmetric version of (FQP), say $(FQP1)$

$(FQP1)$ Find x such that,

$$c^T x + \frac{1}{2} x^T H x \gtrsim Z_0,$$
$$A_i x \lesssim b_i, \ (i = 1, 2, \ldots, m),$$
$$x \geq 0,$$

where Z_0 is the aspiration level of the decision maker. Now taking the tolerances p_i $(i = 0, 1, \ldots, m)$ for the objective and constraint functions and following Zimmermann's [90] approach the membership functions μ_i $(i = 0, 1, \ldots, m)$ are defined as

$$\mu_0(Z) = \begin{cases} 1 & , Z < Z_0, \\ 1 - \dfrac{Z - Z_0}{p_0} & , Z_0 \leq Z \leq Z_0 + p_0, \\ 0 & , Z \geq Z_0 + p_0, \end{cases}$$

and

$$\mu_i(A_i x) = \begin{cases} 1 & , A_i x < b_i, \\ 1 - \dfrac{A_i x - b_i}{p_i} & , b_i \leq A_i x \leq b_i + p_i, \\ 0 & , A_i x > b_i + p_i, \end{cases}$$

Then on the lines of Zimmermann [90], an optimal solution to $(FQP1)$ is obtained by solving

$$\max \quad x_{n+1}$$

subject to,

$$\mu_0(Z) = \left(1 - \frac{Z - Z_0}{p_0}\right) \geq x_{n+1},$$
$$\mu_i(A_i x) = \left(1 - \frac{A_i x - b_i}{p_i}\right) \geq x_{n+1}, \ (i = 1, 2, \ldots, m),$$
$$x_{n+1} \in [0, 1],$$
$$x \geq 0,$$

i.e.

(*SLP*1) max x_{n+1}

subject to,

$$
\begin{aligned}
c^T x + \tfrac{1}{2} x^T H x + p_0 x_{n+1} &\le p_0 + Z_0,\\
A_i x + p_i x_{n+1} &\le p_i + b_i, \quad (i = 1, 2 \ldots m),\\
x_{n+1} &\le 1,\\
x_{n+1} &\ge 0,\\
x &\ge 0.
\end{aligned}
$$

Now clearly (*SLP*1) is of the type (*SLP*) and hence can be solved using Van de Panne's method as described above.

Non-symmetric fuzzy quadratic programming

We now consider the non symmetric version of the fuzzy quadratic programming problem:

(*FQP*2) min $Z = c^T x + \frac{1}{2} x^T H x$

subject to,

$$
\begin{aligned}
A_i x &\lesssim b_i, \ (i = 1, 2, \ldots, m),\\
x &\ge 0.
\end{aligned}
$$

Let p_i $(i = 1, 2, \ldots, m)$ denote the tolerances for the constraint functions. Then following Werners [79], we construct the membership function of the quadratic objective function by defining Z^0 and Z^1 as follows:

$Z^0 = \min \quad c^T x + \frac{1}{2} x^T H x$

subject to,

$$
\begin{aligned}
A_i x &\le b_i, \ (i = 1, 2, \ldots m),\\
x &\ge 0,
\end{aligned}
$$

and,

$Z^1 = \min \quad c^T x + \frac{1}{2} x^T H x$

subject to,

$$
\begin{aligned}
A_i x &\le b_i + p_i, \ (i = 1, 2, \ldots, m),\\
x &\ge 0.
\end{aligned}
$$

Thus, we can construct a continuously nondecreasing linear membership function for the objective by use of Z^0 and Z^1as follows

$$\mu_0(Z) = \begin{cases} 1 & , Z < Z_0, \\ 1 - \dfrac{Z - Z_0}{Z_1 - Z_0} & , Z_0 \le Z \le Z_1, \\ 0 & , Z \ge Z_1. \end{cases}$$

Now using the same membership functions for the constraints as were used for the symmetric approach and along with the membership function of the objective defined above, we proceed in a similar manner as before to get the equivalent crisp programming problem as

$$\begin{aligned} (SLP2) \quad & \max \quad x_{n+1} \\ & \text{subject to,} \\ & c^T x + \tfrac{1}{2} x^T H x + (Z_1 - Z_0) x_{n+1} \le Z_1, \\ & A_i x + p_i x_{n+1} \le p_i + b_i, \ (i = 1, 2, \ldots, m), \\ & x_{n+1} \le 1, \\ & x_{n+1} \ge 0, \\ & x \ge 0, \end{aligned}$$

which is once again of the type (SLP) and hence can be solved using Van de Panne's method.

Example 4.6.1. Consider the fuzzy symmetric quadratic programming problem of the form ($FQP1$) as follows:

Find $(x_1, x_2) \in \mathbb{R}^2$ such that

$$\begin{aligned} 2x_1 + x_2 + 4x_1^2 + 4x_1x_2 + 2x_2^2 &\lesssim 51.88, \\ 4x_1 + 5x_2 &\gtrsim 20, \\ 5x_1 + 4x_2 &\gtrsim 20, \\ x_1 + x_2 &\lesssim 30, \\ x_1, x_2 &\ge 0. \end{aligned}$$

Let the tolerances be given as $p_0 = 2.12$, $p_1 = 2$, $p_2 = 1$, $p_3 = 3$. Then on the lines of ($SLP1$), we have the crisp equivalent of this problem as

$$\begin{aligned} & \max \quad x_3 \\ & \text{subject to,} \\ & 2x_1 + x_2 + 4x_1^2 + 4x_1x_2 + 2x_2^2 + 2.12x_3 \le 54, \\ & 4x_1 + 5x_2 - 2x_3 \ge 18, \\ & 5x_1 + 4x_2 - x_3 \ge 19, \\ & x_1 + x_2 + 3x_3 \le 33, \\ & x_3 \le 1, \\ & x_1, x_2, x_3 \ge 0. \end{aligned}$$

Then in Phase 1, the linear programming problem is as follows

$$\begin{aligned} \max \quad & x_3 \\ \text{subject to,} \quad & \\ & 4x_1 + 5x_2 - 2x_3 \geq 18, \\ & 5x_1 + 4x_2 - x_3 \geq 19, \\ & x_1 + x_2 + 3x_3 \leq 33, \\ & x_3 \leq 1, \\ & x_1, x_2, x_3 \geq 0. \end{aligned}$$

The optimal solution of the above linear programming problem is $x_1^* = 2.22$, $x_2^* = 2.22$, $x_3^* = 1$.

In Phase 2, the quadratic programming problem $(QPP)_\lambda$ that we solve parametrically is as follows

$$\begin{aligned} \min \quad & 2x_1 + x_2 + 4x_1^2 + 4x_1x_2 + 2x_2^2 + 2.12x_3 \\ \text{subject to,} \quad & \\ & 4x_1 + 5x_2 - 2x_3 \geq 18, \\ & 5x_1 + 4x_2 - x_3 \geq 19, \\ & x_1 + x_2 + 3x_3 \leq 33, \\ & x_3 \leq 1, \\ & x_3 \geq \lambda, \\ & x_1, x_2, x_3 \geq 0. \end{aligned}$$

Here we start with $\lambda_0 = c^T x_0 = 1$ and decrease it parametrically as suggested in Van de Panne [73]. An optimal solution of the above quadratic programming problem is $x_1^* = 0.99$, $x_2^* = 3.73$, $x_3^* = 0.86$ and the minimum value of the objective function is 54. Therefore an optimal solution of the given fuzzy quadratic programming problem is $x_1^* = 0.99$, $x_2^* = 3.73$. Also the level of satisfaction of this solution is $x_3^* = 0.86$.

4.7 A two phase approach for solving fuzzy linear programming problems

In the earlier sections we have discussed several methods for solving the fuzzy linear programming problems. These methods are based on the well known Bellman and Zadeh principle and primarily use the max-min operator to obtain a crisp equivalent of the given fuzzy linear programming problem. If the resulting crisp linear programming problem has only one optimal solution then this solution has to be a *fuzzy efficient compromise solution* for the given fuzzy problem. However, if the resulting crisp linear programming problem has multiple optimal solutions then there are examples, e.g. Werners [79] to show that the solution given by the max-min operator may not be efficient but at least one of the multiple optimal solutions is certainly a fuzzy efficient compromise solution.

In this section, we present a two phase approach due to Guu and Wu [21] to take care of the situation where the max-min operator does not produce a fuzzy efficient compromise solution. This method is different from the other two phase methods available in the literature e.g. Lee and Li [46] and Guu and Wu [20]. In this two phase method, the crisp linear programming problem resulting from the max-min operator is solved in Phase I, while in Phase II a solution is obtained which is at least "better" than the solution obtained by the max-min operator. Further the solution resulting by this two phase approach is always a fuzzy efficient compromise solution and that every membership function is at least as large as the one provided by the max-min operator. Thus by using this two phase approach, we not only achieve the highest membership degree in the objective, but also provide for a better utilization of available resources (constraints).

Let us now recall the Werners' approach (Section 4.3) for solving the fuzzy linear programming problem

$$(P1\text{-}FLP1) \qquad \max \quad c^T x$$

$$\text{subject to,}$$

$$\begin{aligned} Ax &\lesssim b, \\ x &\geq 0. \end{aligned}$$

Here we first determine the possible range $[Z_0, Z_1]$ for the objective function of $P1$-$FLP1$ by solving the linear programming problems

$$(LP(b)) \qquad \max \quad c^T x$$

$$\text{subject to,}$$

$$\begin{aligned} Ax &\leq b, \\ x &\geq 0, \end{aligned}$$

and

$$(LP(b+p)) \qquad \max \quad c^T x$$

$$\text{subject to,}$$

$$\begin{aligned} Ax &\leq b + p, \\ x &\geq 0. \end{aligned}$$

Then the membership function of the objective function is taken as

$$\mu_0(c^T x) = \begin{cases} 0 & , c^T x < Z_0, \\ 1 - \left(\dfrac{Z_1 - c^T x}{Z_1 - Z_0} \right) & , Z_0 \le c^T x \le Z_1, \\ 1 & , c^T x > Z_1. \end{cases}$$

Also the membership function of the ith constraint $(i = 1, 2, \ldots, m)$ is chosen as

$$\mu_i(x) = \begin{cases} 1 & , A_i x < b_i \\ 1 - \dfrac{A_i x - b_i}{p_i} & , b_i \le A_i x \le b_i + p_i, \\ 0 & , A_i x > b_i + p_i, \end{cases}$$

where A_i $(i = 1, 2, \ldots, m)$ denotes the ith row of A and p_i denotes the corresponding tolerance level.

Now employing the above membership functions μ_i $(i = 0, 1, \ldots, m)$ and following Bellman and Zadeh principle, the max-min operator results in the following crisp linear programming problem for the problem

(*P*1-*FLP*1) max α

subject to,

$$\begin{aligned} \mu_i(x) &\ge \alpha, \quad (i = 0, 1, 2, \ldots, m), \\ \alpha &\le 1, \\ x &\ge 0, \\ \alpha &\ge 0. \end{aligned}$$

Here we are using the symbol $\mu_i(x)$ uniformly for all i but actually $\mu_0(x)$ refers to $\mu_0(c^T x)$ and, for $i = 1, 2, \ldots, m.$, $\mu_i(x)$ refers to $\mu_i(A_i x)$ as described above.

Let us call the above as Phase I linear programming problem and denote it by (Phase I-*LPP*). Let (x^*, α^*) be an optimal solution of (Phase I-*LPP*). In Phase II, we now construct the linear programming problem, (Phase II-*LPP*) as follows

(Phase II-*LPP*) max $\displaystyle\sum_{i=0}^{m} \alpha_i$

subject to,

$$\begin{aligned}
\mu_i(x) &\geq \alpha_i\,, (i = 0, 1, \ldots, m),\\
\mu_i(x^*) &\leq \alpha_i\,, (i = 0, 1, \ldots, m),\\
\alpha_i &\geq 0, \quad (i = 0, 1, \ldots, m),\\
\alpha_i &\leq 1, \quad (i = 0, 1, \ldots, m),\\
x &\geq 0.
\end{aligned}$$

Let $(x^{**}, \alpha_0^{**}, \alpha_1^{**}, \ldots, \alpha_m^{**})$ be an optimal solution of the problem (Phase II-*LPP*). Also let (*MOP*) denote the multi objective optimization problem,

(*MOP*) $$\max_{x \geq 0} \left(\mu_0(x),\ \mu_1(x), \ldots, \mu_m(x)\right),$$

and call an efficient solution of (*MOP*) as a *fuzzy efficient compromise solution of the problem* (*P*1-*FLP*1). Then we have the following theorem.

Theorem 4.7.1 *The optimal solution x^{**} of the problem (Phase II-LPP) is a fuzzy efficient compromise solution of the problem* (*P*1-*FLP*1).

Proof. If possible, let x^{**} be not a fuzzy efficient compromise solution of (*MOP*). Then there exist a solution $\bar{x}$ of (*MOP*), i.e $\bar{x} \geq 0$, such that

$$\mu_i(x^{**}) \leq \mu_i(\bar{x}), \text{ for all } i = 0, 1 \ldots, m,$$

and

$$\mu_k(x^{**}) < \mu_k(\bar{x}), \text{ for some } k.$$

Since $(x^{**}, \alpha_0^{**}, \alpha_1^{**}, \ldots, \alpha_m^{**})$ is optimal and the coefficient in the objective function of (Phase II-*LPP*) are positive, we have $\alpha_i^{**} = \mu_i(x^{**})$, $(i = 0, 1, \ldots, m)$. Now choosing $\bar{\alpha}_i = \mu_i(\bar{x})$, $(i = 0, 1 \ldots, m)$, we have $(\bar{x},\ \bar{\alpha}_0,\ \bar{\alpha}_1, \ldots, \bar{\alpha}_m)$ feasible for (Phase II-*LPP*) and

$$\sum_{i=0}^{m} \alpha_i^{**} = \sum_{i=0}^{m} \mu_i(x^{**}) < \sum_{i=0}^{m} \mu_i(\bar{x}) = \sum_{i=0,\ i \neq k}^{m} \bar{\alpha}_i + \alpha_k.$$

But this implies that $(x^{**}, \alpha_0^{**}, \alpha_1^{**}, \ldots, \alpha_m^{**})$ is not an optimal solution of (Phase II-*LPP*), which is a contradiction.

Remark 4.7.1. The linear programming problem (Phase II-*LPP*) can be simplified to a somewhat simpler form as follows:

$$\max \quad \sum_{i=0}^{m} \mu_i(x)$$

subject to,

$$\begin{aligned}
\mu_i(x) &\geq \mu_i(x^*), \ (i = 0, 1, \ldots, m),\\
\mu_i(x) &\leq 1, \qquad (i = 0, 1, \ldots, m),\\
x &\in X,
\end{aligned}$$

where X denotes the feasible region of the crisp constraints (if-any) in the original fuzzy problem (*P*1-*FLP*1).

We now illustrate the two phase approach for solving the fuzzy linear programming ($P1$-$FLP1$) with the help of the following example.

Example 4.7.2. (Guu and Wu [21]).) Consider the fuzzy linear programming problem of the type ($P1$-$FLP1$):

$$\begin{array}{ll} \max & 4x_1 + 5x_2 + 9x_3 + 11x_4 \\ \text{subject to,} & \end{array}$$

$$\begin{aligned} x_1 + x_2 + x_3 + x_4 &\lesssim 15, \\ 7x_1 + 5x_2 + 3x_3 + 2x_4 &\lesssim 80, \\ 3x_1 + 4.4x_2 + 10x_3 + 15x_4 &\lesssim 100, \\ x_1, x_2, x_3, x_4 &\geq 0, \end{aligned}$$

with $p_1 = 5, p_2 = 40$, and $p_3 = 30$. Let the objective function be denoted by $Z(x)$ and the three constraint functions be denoted by $g_1(x)$, $g_2(x)$, and $g_3(x)$ respectively. Then for each of the three fuzzy constraints we may take the membership functions as

$$\mu_1(x) = \begin{cases} 1 & , g_1(x) \leq 15, \\ \dfrac{20 - g_1(x)}{5} & , 15 \leq g_1(x) \leq 20, \\ 0 & , g_1(x) > 20, \end{cases}$$

$$\mu_2(x) = \begin{cases} 1 & , g_2(x) < 80, \\ \dfrac{120 - g_2(x)}{40} & , 80 \leq g_2(x) \leq 120, \\ 0 & , g_2(x) > 120, \end{cases}$$

and,

$$\mu_3(x) = \begin{cases} 1 & , g_3(x) < 100, \\ \dfrac{130 - g_3(x)}{30} & , 100 \leq g_3(x) \leq 130, \\ 0 & , g_1(x) > 130. \end{cases}$$

Now, following Werners' approach, we solve the problems $LP(b)$ and $LP(b + p)$ to get $Z_0 = 99.29$ and $Z_1 = 130$. Therefore the membership function of the objective function can be taken as

$$\mu_0(x) = \begin{cases} 1 & , Z(x) > 130, \\ \dfrac{Z(x) - 99.29}{130 - 99.29} & , 99.29 \leq Z(x) \leq 130, \\ 0 & , Z(x) < 99.29. \end{cases}$$

Then using the max-min operator and following Zimmermann's approach we solve the problem (Phase I-LPP)

(Phase I-*LPP*) max α
subject to,

$$\begin{aligned} 4x_1 + 5x_2 + 9x_3 + 11x_4 &\geq 130 - 30.71(1 - \alpha), \\ x_1 + x_2 + x_3 + x_4 &\leq 15 + 5(1 - \alpha), \\ 7x_1 + 5x_2 + 3x_3 + 2x_4 &\leq 80 + 40(1 - \alpha), \\ 3x_1 + 4.4x_2 + 10x_3 + 15x_4 &\leq 100 + 30(1 - \alpha), \\ \alpha &\leq 1, \\ x_1, x_2, x_3, x_4, \alpha &\geq 0. \end{aligned}$$

An optimal solution of the above Phase I problem is $x_1^* = 8.57$, $x_2^* = 0$, $x_3^* = 8.93$, $x_4^* = 0$, $\alpha^* = 0.5$. Further $Z(x^*) = 114.65$, $g_1(x^*) = 17.5$, $g_2(x^*) = 86.78$, $g_3(x^*) = 115.01$. This gives $\mu_0(x^*) = \mu_1(x^*) = \mu_3(x^*) = 0.5$ and $\mu_2(x^*) = 0.83$.
We now construct the (Phase II-*LPP*) given by
(Phase II-*LPP*) max $\alpha_0 + \alpha_1 + \alpha_2 + \alpha_3$
subject to,

$$\begin{aligned} 4x_1 + 5x_2 + 9x_3 + 11x_4 &\geq 130 - 30.71(1 - \alpha_0), \\ x_1 + x_2 + x_3 + x_4 &\leq 15 + 5(1 - \alpha_1), \\ 7x_1 + 5x_2 + 3x_3 + 2x_4 &\leq 80 + 40(1 - \alpha_2), \\ 3x_1 + 4.4x_2 + 10x_3 + 15x_4 &\leq 100 + 30(1 - \alpha_3), \\ 0.5 \leq \alpha_0 &\leq 1, \\ 0.5 \leq \alpha_1 &\leq 1, \\ 0.83 \leq \alpha_0 &\leq 1, \\ 0.5 \leq \alpha_0 &\leq 1, \\ x_1, x_2, x_3, x_4 &\geq 0. \end{aligned}$$

An optimal solution of the above problem is $x_1^{**} = 4.05$, $x_2^{**} = 5.65$, $x_3^{**} = 7.8$, $x_4^{**} = 0$. Also $Z(x^{**}) = Z(x^*) = 114.65$ and $g_1(x^{**}) = 17.5$, $g_2(x^{**}) = 80.00$, $g_3(x^{**}) = 115.01$. This gives $\mu_0(x^{**}) = \mu_1(x^{**}) = \mu_3(x^{**}) = 0.5$ and $\mu_2(x^{**}) = 1$.
Thus by using the two phase approach we get an optimal solution x^{**} which not only achieves the optimal objective value but also gives a higher membership value in μ_2, as $\mu_2(x^{**}) = 1$ and $\mu_2(x^*) = 0.83$.

4.8 Linear goal programming under fuzzy environment

The (crisp) multiobjective linear programming problems can be stated as

$$\begin{array}{ll} \max & (c_1^T x,\ c_2^T x,\ \ldots,\ c_r^T x) \\ \text{subject to,} & \end{array}$$

$$Ax \leq b,$$
$$x \geq 0,$$

where $x \in \mathbb{R}^n$, $b \in \mathbb{R}^m$, $A \in \mathbb{R}^m \times \mathbb{R}^n$, $C \in \mathbb{R}^r \times \mathbb{R}^n$ with c_k^T denoting the kth row of C, $(k = 1, 2, \ldots, r)$. Since there are k objectives to be "maximized", we have also to specify the sense of this "maximization", e.g. *efficiency* or *proper efficiency* etc.

For studying goal programming under fuzzy environment, we prescribe an imprecise aspiration level to each of the p objectives and term these fuzzy objectives as fuzzy goals. Let g_k be the aspiration level assigned to the kth objective $(k = 1, 2, \ldots, r)$. Then a linear goal programming problem under fuzzy environment, in short called fuzzy goal programming and denoted by (*FGP*), can be described as follows:

(*FGP*) Find $x \in \mathbb{R}^n$ so as to satisfy,

$$c_k^T x \gtrsim g_k, \quad (k = 1, 2, \ldots, r),$$

subject to,

$$Ax \leq b,$$
$$x \geq 0.$$

Here the symbol '$\gtrsim$' is to be understood in the fuzzy sense with respect to a chosen membership function. The fuzzy constraints of the form '$\lesssim$' and '$\approxeq$' can be treated similarly.

In the literature there are many approaches to solve the problem (*FGP*) and we wish to discuss some of these in the sequel.

Zimmermann's approach

In this approach, we choose the standard Zimmermann's linear membership function for each of the r fuzzy goals and use the max-min operator to derive the crisp equivalent of the given fuzzy goal programming problem (*FGP*). Specifically, for $k = 1, 2, \ldots, r$, we choose

$$\mu_k(c_k^T x) = \begin{cases} 1 & , c_k^T x \geq g_k, \\ f_k(c_k^T x) = \dfrac{c_k^T x - l_k}{g_k - l_k} & , l_k \leq c_k^T x < g_k, \\ 0 & , c_k^T x < l_k, \end{cases}$$

where l_k is the lower tolerance level for the kth fuzzy goal. If we are using the usual symbol p_k for the tolerance level of the kth fuzzy goal then $l_k = g_k - p_k$.

Now using the max-min operator, Zimmermann's model [90] for the problem (*FGP*) is obtained as

(*ZLP*)
$$\begin{aligned} \max \quad & \lambda \\ \text{subject to,} \quad & \\ & \lambda \leq \frac{c_k^T x - l_k}{g_k - l_k}, \ (k = 1, 2, \ldots, r), \\ & Ax \leq b, \\ & \lambda \leq 1, \\ & x, \lambda \geq 0. \end{aligned}$$

Here $f_k(c_k^T x) = \dfrac{c_k^T x - l_k}{g_k - l_k}$ is linear, but in general it can be piecewise linear that is concave or quasi concave. In the literature there are extensions of Zimmermann's model for the case when $f_k(c_k^T x)$ is concave or quasi-concave e.g. Narsimhan ([58], [59]) and Hannan ([23], [24]) for the concave case, and Nakumura [57] and Yang et al. [88], Inuiguchi et al. [25], Li and Yu [43], Wang and Fu [78], and Lin and Chen [47] for the quasi concave case.

An immediate extension of Zimmermann's approach is the two phase approach of Li [44]. In Section 4.7, we have already learnt a two phase approach for the case when there is a single objective only. Here we describe a similar approach for the multiobjective scenario. In this approach, the first phase utilizes Zimmermann's approach and solves the resulting crisp problem (*ZLP*) as described above. Let $(x^*, \ \lambda^*)$ be an optimal solution of (*ZLP*). If x^* is unique, then it is taken as an optimal solution of the fuzzy goal programming problem (*FGP*), otherwise, in Phase II, the following linear programming problem is formulated:

$$\begin{aligned} \max \quad & \sum_{k=1}^{r} \frac{\lambda_k}{r} \\ \text{subject to,} \quad & \\ & \lambda_k^* \leq \lambda_k \leq f_k(c_k^T x), \ (k = 1, 2, \ldots, r), \\ & Ax \leq b, \\ & x \geq 0. \end{aligned}$$

If $(\bar{x}, \ \bar{\lambda})$ is an optimal solution of the above linear programming problem then $\bar{x}$ is an efficient solution of the multiobjective linear programming problem (*MOP*) where

$$\begin{aligned} \max \quad & (\mu_1(x), \mu_2(x), \ldots, \mu_r(x)) \\ \text{subject to,} \quad & \\ & Ax \leq b, \\ & x \geq 0, \end{aligned}$$

where, as before, $\mu_k(x)$ actually stands for $\mu_k(c_k^T x)$ for $k = 1, 2, \ldots, r$.

Weighted max-min approach

In the context of fuzzy goal programming, it is important to note that in actual applications the objectives have different level of importance and therefore are not equally preferable. In such a situation, the Zimmermann's approach ([90], [91]) as described above is not suitable and a weighted max-min approach seems to be very natural. Keeping this in mind Lai and Hwang [36] proposed the following problem to solve the given fuzzy goal programming problem (*FGP*)

$$\begin{aligned} \max \quad & \lambda + \delta \sum_{k=1}^{r} w_k f_k(c_k^T x) \\ \text{subject to,} \quad & \\ & \lambda \leq f_k(c_k^T x), \ (k = 1, 2, \ldots, r), \\ & Ax \leq b, \\ & x \geq 0. \end{aligned}$$

Here w_k is the relative weight of the kth objective and δ is a sufficiently small positive number. Also $e^T w = 1$, where $e^T = (1, 1, \ldots, 1) \in \mathbb{R}^r$.

Though the above model takes into consideration the relative weights of the objectives, it does not give importance to objectives with higher weights as δ is a very small positive number. In fact this model of Lai and Hwang [36] will give the same solution as that of Zimmermann's problem (*ZLP*). To counter this difficulty, Lin [45] presented another weighted max-min model which is based on the logic that when the decision maker provides relative weights for fuzzy goals together with appropriate membership functions, the ratio of the achieved levels should be close to the ratio of the objective weights as best as possible. This aim can be achieved by solving the following weighted Zimmermann's type linear programming problem (*WZLP*).

$$\begin{aligned} (WZLP) \quad \max \quad & \lambda \\ \text{subject to,} \quad & \end{aligned}$$

$$\begin{aligned} w_k\lambda &\le f_k(c_k^T x), (k = 1, 2, \ldots, r), \\ Ax &\le b, \\ x, \lambda &\ge 0. \end{aligned}$$

Here it may be noted that in the above problem there is no condition that $\lambda \le 1$. In fact λ can be more than unity because $w_k < 1$. But the actual achieved level for each objective will never exceed unity, which can be computed by solving the above problem to get (x^*, λ^*) and then utilizing the definition of the membership function $\mu_k(c_k^T x)$, for $k = 1, 2, \ldots, r$.

We now try to argue that the above weighted max-min model achieves the objective of obtaining an optimal solution so that the ratio of the achieved levels of objectives is close to the ratio of the weights as best as possible. For this, let s_k denote the slack variable for the kth constraint $w_k\lambda \le \mu_k(c_k^T x)$, i.e. for $k = 1, 2, \ldots, r$, we have

$$w_k\lambda + s_k \le \mu_k(c_k^T x).$$

Also let us recollect that while searching for an optimal solution of the linear programming problem we tend to use as many recourses as possible, i.e. we tend to reduce the slack and surplus variables as best as possible. Since the membership functions $\mu_k(c_k^T x)$, $(k = 1, 2, \ldots, r)$, are bounded quasi concave functions, the problem $(WZLP)$ can not have unbounded solutions. Therefore in $(WZLP)$, as λ is maximized the slack variables s_k $(k = 1, 2, \ldots, r)$ are minimized, i.e. the constraint $\lambda w_k \le \mu_k(c_k^T x)$ becomes close to an equality. In other words, the achieved level of the kth objective $\mu_k(c_k^T x)$ becomes close to λw_k as best as possible. In ideal case, all slack variables s_k will be zero and so the ratio of achieved levels will be same as the ratio of the weights.

The following example illustrates the above discussion:

Example 4.8.1. (Lin [45], Hannan [23]).

Find $x \in \mathbb{R}^3$ to satisfy,

$$\begin{aligned} Z_1 &= 3x_1 + x_2 + x_3 \gtrsim 7, \\ Z_2 &= x_1 - x_2 + 2x_3 \gtrsim 8, \\ Z_3 &= x_1 + 2x_2 \qquad \gtrsim 5, \end{aligned}$$

subject to,

$$\begin{aligned} 4x_1 + 2x_2 + 3x_3 &\le 10, \\ x_1 + 3x_2 + 2x_3 &\le 8, \\ x_3 &\le 5, \\ x_1, x_2, x_3 &\ge 0. \end{aligned}$$

Let the membership functions for the three goals be given by the following concave piecewise linear membership functions:

$$\mu_1(Z_1) = \begin{cases} 1 & , Z_1 \geq 7, \\ 0.2(Z_1 - 6) + 0.8 & , 6 \leq Z_1 < 7, \\ 0.3(Z_1 - 5) + 0.5 & , 5 \leq Z_1 < 6, \\ 0.5(Z_1 - 4) & , 4 \leq Z_1 < 5, \\ 0 & , Z_1 < 4, \end{cases}$$

$$\mu_2(Z_2) = \begin{cases} 1 & , Z_2 \geq 8, \\ 0.15(Z_2 - 4) + 0.4 & , 4 \leq Z_2 < 8, \\ 0.2(Z_2 - 2) & , 2 \leq Z_2 < 4, \\ 0 & , Z_2 < 2, \end{cases}$$

$$\mu_1(Z_3) = \begin{cases} 1 & , Z_3 \geq 5, \\ 0.2(Z_3 - 4) + 0.8 & , 4 \leq Z_3 < 5, \\ 0.4(Z_3 - 2) & , 2 \leq Z_3 < 4, \\ 0 & , Z_3 < 2, \end{cases}$$

Also, let the decision maker prescribe the relative weights of the three objective as $w_1 = 0.4$, $w_2 = 0.35$ and $w_3 = 0.25$.
Now from the given data, and following Yang et al. [88] the weighted max-min model leads to the following weighted Zimmermann type linear programming problem (*WZLP*):

$$\begin{aligned} \max \quad & \lambda \\ \text{subject to,} \quad & \\ & 0.4\lambda \leq 0.2(Z_1 - 6) + 0.8, \\ & 0.4\lambda \leq 0.3(Z_1 - 5) + 0.5, \\ & 0.4\lambda \leq 0.5(Z_1 - 4), \\ & 0.35\lambda \leq 0.15(Z_2 - 4) + 0.4, \\ & 0.35\lambda \leq 0.2(Z_2 - 2), \\ & 0.25\lambda \leq 0.2(Z_3 - 4) + 0.8, \\ & 0.25\lambda \leq 0.4(Z_3 - 2), \\ & 4x_1 + 2x_2 + 3x_3 \leq 10, \\ & x_1 + 3x_2 + 2x_3 \leq 8, \\ & x_3 \leq 5, \\ & x_1, x_2, x_3 \geq 0. \end{aligned}$$

where $Z_1 = 3x_1 + x_2 + x_3$, $Z_2 = x_1 - x_2 + 2x_3$, $Z_3 = x_1 + 2x_2$.
The above linear programming problem has an optimal solution $\lambda^* =$

0.82, $x_1^* = 0.60$, $x_2^* = 0.95$ and $x_3^* = 1.89$. For this solution $Z_1 = 4.656$, $Z_2 = 3.435$ and $Z_3 = 2.513$. Also the achieved levels of goals are $\mu_1 = 0.328$, $\mu_2 = 0.287$ and $\mu_3 = 0.205$. It can be verified that

$$\frac{\mu_1}{w_1} = \frac{0.328}{0.4} = 0.82,$$

$$\frac{\mu_2}{w_2} = \frac{0.287}{0.35} = 0.82,$$

and

$$\frac{\mu_3}{w_3} = \frac{0.205}{0.25} = 0.82,$$

where $\lambda^* = 0.82$ is the optimal achieved level of the problem ($WZLP$).

If this problem is also solved by the Zimmermann's approach, i.e. by solving the problem (ZLP), we obtain $x_1^* = 0.50$, $x_2^* = 1.08$, $x_3^* = 1.95$, $\lambda^* = 0.789$. This gives $Z_1 = 4.526$, $Z_2 = 3.315$, $Z_3 = 2.658$, $\mu_1 = 0.263 = \mu_2 = \mu_3$. Here it may be noted that in (ZLP) we do not assign weights to different objectives i.e. take all objectives equally important. Therefore in ($WZLP$) if we take $w_1 = w_2 = w_3$, we shall get the same solution as given by the Zimmermann's problem (ZLP). In fact the problem of Lai and Hwang [36] will also give the same solution as that of (ZLP) because $\delta > 0$ is small.

Comparing these two solutions, i.e. the solutions of ($WZLP$) and (ZLP), we observe that the weighted max-min approach of Lin [45] as described above, finds an optimal solution that gives the objective function Z_1 a higher level of achievement (0.328) in comparison to the other two objectives . This is because it has been given more weight than the other two objectives. Also this solution has the property that the ratio of achieved levels is the same as the ratio of the weights of objectives.

A crisp goal programming approach

Mohamed [54] recently proposed a different approach to solve the fuzzy goal programming problem (FGP). This approach is based on various models for solving the crisp linear goal programming problem (e.g. Taha [72]), and is based on the argument that as the maximum value of any membership function $\mu_k(c_k^T x)$ in the problem (ZLP) is 1, maximizing any of these is equivalent to making it close to 1 as best as possible. This can be achieved by minimizing its negative deviational variable d_k^- from 1. Thus here only the negative (under) deviation variable d_k^- is required to be minimized to achieve the aspired levels of the fuzzy goals. In this context it may be noted that any other (positive) deviation from a fuzzy goal indicates the full achievement of the membership value as the value can not be more than 1.

We know that for $k = 1, 2, \ldots, r$, the Zimmermann's type membership function $\mu_k(c_k^T x)$ of the fuzzy goal $Z_k = c_k^T x \gtrsim g_k$ is given by

$$\mu_k(c_k^T x) = \begin{cases} 1 & , c_k^T x \geq g_k, \\ \dfrac{c_k^T x - l_k}{g_k - l_k} & , l_k \leq c_k^T x < g_k, \\ 0 & , c_k^T x < l_k. \end{cases}$$

Since we wish to achieve the aspired levels of the fuzzy goals, we have to keep $\mu_k(c_k^T x)$, $(k = 1, 2, \ldots, r)$, close to 1 as best as possible. Thus we have to minimize the under deviational variable d_k^- for each of the k fuzzy goals. This requirement we express in the form of the following constraints

$$\begin{aligned} \frac{Z_k(x) - l_k}{g_k - l_k} + d_k^- - d_k^+ &= 1, \\ d_k^- . d_k^+ &= 0, \\ d_k^- &\geq 0, \\ d_k^+ &\geq 0, \end{aligned}$$

and introduce a suitable measure of negative deviations d_k^- so that d_k^- is minimized for all $k = 1, 2, \ldots, r$.

In conventional (crisp) goal programming problem *GP*, the under and/or over deviational variables d_k^- and d_k^+ are included in the achievement function (objective function) for minimizing them but here, as explained, only d_k^- will be included. Since there are many choices of the objective function for the (crisp) linear goal programming problem we have similar formulations for the fuzzy goal programming problem as well. In particular, if we choose the *minsum GP* as the model, then the equivalent optimization problem for the given fuzzy goal programming problem (*FGP*) is given by

$$\min \quad \sum_{k=1}^{r} w_k^- d_k^-$$

subject to,

$$\begin{aligned} \frac{Z_k(x) - l_k}{g_k - l_k} + d_k^- - d_k^+ &= 1, \ (k = 1, 2, \ldots, r), \\ Ax &\leq b, \\ d_k^- . d_k^+ &= 0, \ (k = 1, 2, \ldots, r), \\ x, d_k^-, d_k^+ &\geq 0. \end{aligned}$$

But, for $(k = 1,2,\ldots,r)$, $Z_k(x) = c_k^T x$ and hence we have

$$\frac{c_k^T x - l_k}{g_k - l_k} + d_k^- - d_k^+ = 1,$$

i.e. $c_k^T x + (g_k - l_k)d_k^- - (g_k - l_k)d_k^+ = g_k$.

Therefore the above *minsum GP* becomes

$$\begin{aligned} \min \quad & \sum_{k=1}^{r} w_k^- d_k^- \\ \text{subject to,} \quad & \\ & c_k^T x + p_k d_k^- - p_k d_k^+ = g_k, \ (k = 1,2,\ldots,r), \\ & Ax \leq b, \\ & d_k^- . d_k^+ = 0, \ \ (k = 1,2,\ldots,r), \\ & x, d_k^-, d_k^+ \geq 0. \end{aligned}$$

Here $p_k = g_k - l_k$ is the admissible violation for the kth goal.

To solve the above optimization problem we have to choose the weights w_k^- for the variable d_k^-. These weights can either be chosen subjectively or as suggested by Mohamed [54] can be chosen as $w_k^- = \dfrac{1}{g_k - l_k} = \dfrac{1}{p_k}$. This crisp formulation of the given fuzzy goal programming (*FGP*) is not unique as we can choose other objective functions corresponding to linear goal programming, e.g. the max-min, lexicographical form etc. In the min-max form the objective is to minimize the maximum of d_k^-, $(k = 1,2,\ldots,r)$. Therefore if we take $u = \max_k (d_k^-)$, then the optimization problem to be solved is

$$\begin{aligned} \min \quad & u \\ \text{subject to,} \quad & \\ & c_k^T x + p_k d_k^- - p_k d_k^+ = g_k, \ (k = 1,2,\ldots,r), \\ & u \geq d_k^-, \ (k = 1,2,\ldots,r), \\ & Ax \leq b, \\ & d_k^- . d_k^+ \geq 0, \ \ (k = 1,2,\ldots,r), \\ & x, d_k^-, d_k^+ \geq 0. \end{aligned}$$

Example 4.8.2. Consider the following fuzzy goal programming problem (*FGP*):

$$\text{Find } (x_1, x_2) \in \mathbb{R}^2 \text{ to satisfy}$$

$$\begin{aligned} Z_1 &= -x_1 + 2x_2 \gtrsim 14, \\ Z_2 &= 2x_1 + 3x_2 \gtrsim 21, \end{aligned}$$

subject to,

$$\begin{aligned} -x_1 + 3x_2 &\leq 21, \\ x_1 + 3x_2 &\leq 27, \\ 4x_1 + 3x_2 &\leq 45, \\ 3x_1 + x_2 &\leq 30, \\ x_1, x_2 &\geq 0. \end{aligned}$$

Here $g_1 = 14$ and $g_2 = 21$. Let $p_1 = 8$ and $p_2 = 10$. Then to solve the above fuzzy goal programming problem, using the Zimmermannn's approach, we have to solve the problem (*ZLP*) given below.

(*ZLP*) max λ

subject to,

$$\begin{aligned} \lambda &\leq 1 - \frac{(14 + x_1 - 2x_2)}{8}, \\ \lambda &\leq 1 - \frac{(21 - 2x_1 - x_2)}{10}, \\ -x_1 + 3x_2 &\leq 21, \\ x_1 + 3x_2 &\leq 27, \\ 4x_1 + 3x_2 &\leq 45, \\ 3x_1 + x_2 &\leq 30, \\ x_1, x_2, \lambda &\geq 0. \end{aligned}$$

It can be verified that an optimal solution of (*ZLP*) is $x_1^* = 4.53$, $x_2^* = 7.49$ and $\lambda^* = 0.56$.

The given fuzzy linear programming problem can also be solved by using the Mohamed's approach. In particular we can take the min-max form of the (crisp) linear goal programming problem as,

min u

subject to,

$$\begin{aligned} -x_1 + 2x_2 + 8d_1^- - 8d_1^+ &= 14, \\ 2x_1 + x_2 + 10d_2^- - 10d_2^+ &= 21, \\ u &\geq d_1^-, \\ u &\geq d_2^-, \\ -x_1 + 3x_2 &\leq 21, \\ x_1 + 3x_2 &\leq 27, \\ 4x_1 + 3x_2 &\leq 45, \\ 3x_1 + x_2 &\leq 30, \\ d_1^-.d_1^+ &= 0, \\ d_2^-.d_2^+ &= 0, \\ x_1, x_2, d_1^-, d_1^+, d_2^-, d_2^+ &\geq 0. \end{aligned}$$

The above problem can be solved to get $x_1^* = 4.53$, $x_2^* = 7.49$ and $u^* = 0.44$. It is not surprising to note that this solution is the same as the one obtained by solving (ZLP) where $u^* = (1 - \lambda^*)$. This is because of an equivalence theorem established by Mohamed [54].
In case we are solving the minsum form of the (crisp) linear programming problem, then the corresponding problem for the given fuzzy goal programming problem is

$$\min \quad \frac{1}{8}d_1^- + \frac{1}{10}d_2^-$$

subject to,

$$\begin{aligned}
-x_1 + 2x_2 + 8d_1^- - 8d_1^+ &= 14,\\
2x_1 + x_2 + 10d_2^- - 10d_2^+ &= 21,\\
-x_1 + 3x_2 &\leq 21,\\
x_1 + 3x_2 &\leq 27,\\
4x_1 + 3x_2 &\leq 45,\\
3x_1 + x_2 &\leq 30,\\
d_1^-.d_1^+ &= 0,\\
d_2^-.d_2^+ &= 0,\\
x_1, x_2, d_1^-, d_1^+, d_2^-, d_2^+ &\geq 0.
\end{aligned}$$

Remark 4.8.3. Pal et al. [63] have recently extended the (crisp) linear goal programming approach of Mohamed [54] for solving fuzzy goal programming problems with linear fractional objectives. Although for the linear fractional case the constraint

$$\frac{Z_k(x) - l_k}{g_k - l_k} + d_k^- - d_k^+ = 1,$$

is inherently nonlinear, as $Z_k(x) = \dfrac{c_k^T x + \alpha_k}{h_k^T x + \beta_k}$ with $h_k^T x + \beta_k > 0$, a suitable linearization procedure can be carried out to express the above constraint as

$$H_k^T x + D_k^- - D_k^+ = G_k.$$

Here $D_k^- = (h_k^T x + \beta_k)d_k^-$, $D_k^+ = (h_k^T x + \beta_k)d_k^+$, $G_k = L_k' \beta_k - L_k \alpha_k$, $H_k = L_k c_k - L_k' h_k$, $L_k = \dfrac{1}{g_k - l_k}$ and $L_k' = 1 + L_k l_k$. Also as $d_k^- \leq 1$, there will be an additional constraint of the form $\dfrac{D_k^-}{h_k^T x + \beta_k} \leq 1$ i.e. $-h_k^T x + D_k^- \leq \beta_k$. There is of course no such condition required for d_k^+.

4.9 Conclusions

In this chapter we have discussed various models for studying fuzzy linear programming problems. Here no attempt has been made to be exhaustive but certainly some of the most basic models available in the literature have been included. Some other important models of fuzzy decision making, e.g. modality constrained programming and fuzzy (valued) relations approach to fuzzy linear programming are not discussed here as these are included in Chapter10. As far as multiobjective linear programming in fuzzy environment is concerned, we have discussed only the goal programming approach and have not even attempted to discuss other approaches. An appropriate references for this chapter is the excellent book by Lai and Hwang [37] which has many other references for single as well as multiobjective fuzzy linear programming problems. Further a detailed study of fuzzy multiobjective decision making is available in Lai and Hwang [38].

5

Duality in linear and quadratic programming under fuzzy environment

5.1 Introduction

In Chapter 1, we have already seen the important role played by the linear programming duality in the context of (crisp) two person zero-sum matrix game theory. Realizing this importance, it is imperative to look for "similar" duality results for linear programming in fuzzy environment so as to use them for solving fuzzy matrix games.

This chapter discusses some of the most basic models available in the literature for studying duality in fuzzy linear programming while some recent ones will be presented in Chapter 7 and Chapter 10 where they are probably more relevant. In this context, it may be noted that though there is a vast literature on modeling and solution procedures in fuzzy linear programming the studies on duality have been rather scarce until very recently.

This chapter consists of four main sections, namely, *duality in linear programming under fuzzy environment: Rödder and Zimmermann's model, a modified linear programming duality under fuzzy environment, Verdegay's dual for fuzzy linear programming, and duality for quadratic programming under fuzzy environment.*

5.2 Duality in LP under fuzzy environment: Rödder-Zimmermann's model

It is known that the classical linear programming duality can be understood in terms of the maxmin and the minmax problems of the associated Lagrangian function $L(x, u)$. Specifically, for the linear programming problem

$$(LP) \qquad \max \quad c^T x$$
$$\text{subject to,}$$
$$Ax \leq b,$$
$$x \geq 0,$$

we associate the Lagrangian $L : \mathbb{R}^n_+ \times \mathbb{R}^m_+ \to \mathbb{R}$ given by $L(x,u) = c^T x + u^T(b - Ax)$ and obtain (LP) and its dual (LD) as "$\max_{x\geq 0} \min_{u \geq 0} L(x,u)$" and "$\min_{u\geq 0} \max_{x \geq 0} L(x,u)$" respectively. This later problem can further be simplified to get

$$(LD) \qquad \min \quad b^T u$$
$$\text{subject to,}$$
$$A^T u \geq c,$$
$$u \geq 0.$$

Here $x \in \mathbb{R}^n$, $c \in \mathbb{R}^n$, $b \in \mathbb{R}^m$, $u \in \mathbb{R}^m$ and $A \in \mathbb{R}^m \times \mathbb{R}^n$.

Rödder and Zimmermann [67] generalized these classical maxmin and minmax problems, which are associated with $L(x,u)$, to take into consideration the fuzzy environment. In the crisp case for every decision x of the primal (say, Industry and denote it by (I)) there exists a decision u of the dual (say, Market and denote it by (M)) and vice versa. When we fuzzify the case, there is a fuzzy set on the solution space X, that is to each x in X there corresponds a grade of membership which represents the satisfaction of the decision maker (I) with the decision x. Also corresponding to each x in X there exists a fuzzy set on U, that is to each x in X there exists a grade of membership for each decision u which represents the satisfaction of the decision maker (I) with the decision u. Thus we have a fuzzy set $\left\{(x,\ \mu^I(x)) :\ x \in X\right\}$ on X and a family of fuzzy sets $\left\{(u,\ \mu^I_x(u)) :\ u \in U\right\}$ on U with the parameter $x \in X$. Specifically in our case, $\mu^I(x)$ is the membership function of (I) on $\{x :\ x \geq 0\}$ and $\mu^I_x(u)$ is the membership function of (I) on $\{u :\ u \geq 0\}$ for any given $x \geq 0$. Now we define another fuzzy set on U, called *mixture* of μ^I and μ^I_x, having the membership function as

$$v^I(u) = \max_{x \geq 0} \min\,(\mu^I(x),\ \mu^I_x(u)).$$

This "mixture", having membership function $v^I(u)$, demands (I) to determine a family of decisions $x(u)$ such that for each decision $u \geq 0$ of the competitor (M), the optimum of $\mu^I(x)$ and $\mu^I_x(u)$ is reached.

Similarly, if $\mu^M(u)$ is the membership function of (M) on $\{u :\ u \geq 0\}$ and $\mu^M_u(x)$ is the membership function of M on $\{x :\ x \geq 0\}$ for any given

$u \geq 0$, then for (M), we have to consider the mixture of fuzzy sets μ^M and μ_u^M and get

$$\xi^M(x) = \max_{u \geq 0} \min\left(\mu^M(x),\ \mu_u^M(x)\right).$$

Rödder and Zimmermann [67] suggested the following choices for the functions μ^I, μ_x^I, μ^M, μ_u^M

$$\mu^I(x) = \begin{cases} 1 & , c^T x^0 \leq c^T x, \\ 1 - (c^T x^0 - c^T x) & , \text{otherwise}, \end{cases}$$

$$\mu_x^I(u) = \begin{cases} 0 & , u^T(b - Ax) \leq 0, \\ u^T(b - Ax) & , \text{otherwise}, \end{cases}$$

$$\mu^M(u) = \begin{cases} 1 & , b^T u \leq b^T u^0, \\ 1 - (b^T u - b^T u^0) & , \text{otherwise}, \end{cases}$$

and

$$\mu_u^M(x) = \begin{cases} 0 & , x^T(c - A^T u) \leq 0, \\ -x^T(c - A^T u) & , \text{otherwise}. \end{cases}$$

Here $c^T x^0$ is the aspiration level of (I) and $b^T u^0$ is the aspiration level of (M). Substituting the membership functions μ^I, μ_x^I, μ^M, μ_u^M as above, we obtain the following linear programming problems for (I) and (M) respectively

$(FP)_u$ max λ_1

subject to,

$$\begin{aligned} &\lambda_1 \leq \left(1 - (c^T x^0 - c^T x)\right), \\ &\lambda_1 \leq u^T(b - Ax), \quad (\text{for any given } u \geq 0), \\ &x \geq 0, \end{aligned}$$

$(FD)_x$ min $(-\lambda_2)$

subject to,

$$\begin{aligned} &\lambda_2 \geq \left((b^T u - b^T u^0) - 1\right), \\ &\lambda_2 \geq x^T(c - A^T u), \quad (\text{for any given } x \geq 0), \\ &u \geq 0. \end{aligned}$$

Here $\lambda_1 = \min\ (\mu^I(x),\ \mu^I_x(u))$ and $(-\lambda_2) = \eta = \min\ (\mu^M(u),\ \mu^M_u(x))$. Also depending upon $u \geq 0$ (respectively $x \geq 0$), we have a problem of type *(FP)* (respectively *(FD)*), and therefore to emphasize this point and make it more explicit we have denoted these problems as $(FP)_u$ and $(FD)_x$ respectively.

The above pair $(FP)_u$-$(FD)_x$ is called the fuzzy primal-dual pair of linear programming problems.

Lemma 5.2.1. *If there exist $\hat{x}$ feasible for $(FP)_u$ such that $c^T\hat{x} > 0$ and $-u^TA\hat{x} > 0$, then $(FP)_u$ has an unbounded solution.*

Proof Since $\hat{x}$ is feasible for $(FP)_u$ we have $\lambda_1 \leq (1 - c^Tx^0) + c^T\hat{x}$ and $\lambda_1 \leq u^T(b - A\hat{x})$: As $c^T\hat{x} > 0$, we can choose $\epsilon > 0$ large enough and have $\lambda_1 \leq (1 - c^Tx^0) + c^T(\epsilon\hat{x})$. Similarly, we also have $\lambda_1 \leq u^Tb - u^TA\hat{x} \leq u^Tb - u^TA(\epsilon\hat{x})$ as $-u^TA\hat{x} > 0$. Therefore for $\epsilon > 0$, large enough, $(\epsilon\hat{x}, \lambda_1)$ is feasible for $(FP)_u$. Since $\epsilon > 0$ can be made arbitrarily large, λ_1 can be allowed to take any arbitrarily large value and so $(FP)_u$ has unbounded solution.

Definition 5.2.1 (A set of reasonable decisions). *The set $U_0 = \{\, u \geq 0 :\ \nexists\, x \geq 0$ such that $c^Tx > 0$ and $-u^TAx > 0 \,\}$ is called the set of reasonable decisions u for (M).*

In view of Lemma 5.2.1, a decision u for *(M)* should always be in U_0. Otherwise *(I)* can increase the value of its membership function arbitrary; and therefore the word "reasonable" is very appropriate. Similarly the set $X_0 = \{\, x \geq 0 :\ \nexists\, u \geq 0$ such that $b^Tu < 0$ and $-u^TAx < 0 \,\}$ is called the set of reasonable decisions x for *(I)*, and x must always be in X_0 otherwise *(M)* can decrease the value of its membership function arbitrary.

We now consider two more sets U_1 and X_1 given by

$$U_1 = \{\, u \geq 0 : \forall\, x \geq 0,\ c^Tx \geq 0 \ \Rightarrow\ u^TAx \geq 0 \,\},$$

and

$$X_1 = \{\, x \geq 0 : \forall\, u \geq 0,\ b^Tu \leq 0 \ \Rightarrow\ u^TAx \leq 0 \,\}.$$

Lemma 5.2.2. *For the sets U_1 and X_1 we have $U_1 \subset U_0$ and $X_1 \subset X_0$, the containment being proper.*

Proof. The proof follows by noting that the sets U_0 and X_0 can also be expressed as

$$U_0 = \{\, u \geq 0 : \forall\, x \geq 0,\ c^Tx > 0 \ \Rightarrow\ u^TAx \geq 0 \,\}$$

and,

$$X_0 = \{\, x \geq 0 : \forall\, u \geq 0,\ b^T u < 0 \Rightarrow u^T A x \leq 0 \,\}.$$

Now it can be argued that by restricting the decisions u to U_1 and decisions x to X_1, we do not loose any economically relevant information. This is because the only additional condition U_1 has with respect to U_0 is that, for $x \geq 0$, $c^T x = 0 \Rightarrow u^T A x \geq 0$. But such a decision u of (M) does not make $(FP)_u$ unbounded and so it does not allow (I) to increase its membership value arbitrarily. Similar arguments hold for the choice of X_1 as well.

Next, we aim to characterize the sets U_1 and X_1. For thus we use the well known Farkas lemma (Mangasarian [53]).

Lemma 5.2.3. (Farkas lemma). *Let A be a given $(m \times n)$ real matrix and b be a given n vector. The inequality $b^T y \geq 0$ holds for all vectors $y \in \mathbb{R}^n$ satisfying $Ay \geq 0$ if and only if there exists an m vector ρ, $\rho \geq 0$ such that $A^T \rho = b$.*

Lemma 5.2.4. *The sets U_1 and X_1 have the following representation:*

$$U_1 = \{\, u \geq 0 :\ \exists\, \alpha \in \mathbb{R},\ \alpha \geq 0 \text{ such that } \alpha c \leq A^T u \,\}$$

and,

$$X_1 = \{\, x \geq 0 :\ \exists\, \beta \in \mathbb{R},\ \beta \geq 0 \text{ such that } Ax \leq \beta b \,\}.$$

Proof. Let $u \in U_1$. Then $x \geq 0$, $c^T x \geq 0$ implies $(u^T A) x \geq 0$. Hence by Farkas lemma, there exists an $(n+1)$ vector ρ, $\rho \geq 0$ such that

$$\begin{pmatrix} 1 \\ c^T \end{pmatrix}^T \rho = (u^T A)^T.$$

Let $\rho = \begin{pmatrix} \beta_n \\ \alpha \end{pmatrix}$. Then we have

$$\begin{pmatrix} 1 & c \end{pmatrix} \begin{pmatrix} \beta_n \\ \alpha \end{pmatrix} = A^T u,$$

i.e.

$$\beta_n + c\alpha = A^T u.$$

Since $\rho \geq 0$, we have $\beta_n \geq 0$ and thus $A^T u \geq c\alpha$. The proof for X_1 is similar.

In view of above discussion, it makes sense to call following two problems as fuzzy primal-dual pair of linear programming problems and attempt to establish possible duality relations for them.

$$
\begin{aligned}
(FP1)_u \quad & \max \quad \lambda_1 \\
& \text{subject to,} \\
& \lambda_1 \leq 1 - (c^T x^0 - c^T x), \\
& \lambda_1 \leq u^T(b - Ax), \quad \text{(for any given } u \geq 0\text{)}, \\
& \alpha c \leq A^T u, \\
& Ax \leq \beta b, \\
& x, \alpha, \beta \geq 0,
\end{aligned}
$$

and

$$
\begin{aligned}
(FD1)_x \quad & \min \quad (-\lambda_2) \\
& \text{subject to,} \\
& \lambda_2 \geq (b^T u - b^T u^0) - 1, \\
& \lambda_2 \geq x^T(c - A^T u), \quad \text{(for any given } x \geq 0\text{)}, \\
& A^T u \geq \alpha c, \\
& Ax \leq \beta b, \\
& u, \alpha, \beta \geq 0.
\end{aligned}
$$

For this, we note from the definition of sets U_1 and X_1 that for every $(x, u) \in X_1 \times U_1$, there exist $\alpha, \beta \in \mathbb{R}_+$ such that $\alpha c \leq A^T u$ and $Ax \leq \beta b$. This gives $\alpha(c^T x) \leq u^T Ax \leq \beta(b^T u)$. Hence for the optimal solutions of $(FP1)_u$ and $(FD1)_x$ we have

$$\alpha\left(c^T x_{opt}(u)\right) \leq \beta(u)(b^T u),$$

and,

$$\alpha(x)(c^T x) \leq \beta\left(b^T u_{opt}(x)\right).$$

Here we should note that as per the definition of sets X_1 and U_1, α depends on x while β depend on u.

Now for the economically meaningful cases, α and β are non-zero and therefore the above inequalities give

$$\frac{\alpha}{\beta(u)}\left(c^T x_{opt}(u)\right) \leq (b^T u),$$

and,

$$c^T(x) \leq \frac{\beta}{\alpha(x)}\left(b^T u_{opt}(x)\right),$$

which are the generalizations of the usual (crisp) weak duality theorem.

Remark 5.2.5. In the classical duality theory of linear programming, the optimal values of both, the primal and the dual, linear programming problems are same. For duality in fuzzy linear programming, since violations are permitted and the criterion being the fulfilling of certain aspiration levels, it should not be expected that the optimal values of the fuzzy primal-dual pair should be equal.

5.3 A modified linear programming duality under fuzzy environment

In the above construction of Rödder and Zimmermann, the membership functions $\mu^I(x)$, $\mu^M(u)$ take values in $(-\infty, 1]$, while $\mu^I_x(u)$ and $\mu^M_u(x)$ take values in $[0, \infty)$. This is not in conformity with the usual practice that the membership function should take values in [0,1]. Further, if $\lambda_1 = 1$ and $-\lambda_2 = 1$ in $(FP)_u$ and $(FD)_x$, respectively, then $u^T(b-Ax) \geq 1$ and $x^T(c - A^T u) \leq -1$, whereas in the crisp scenario one should have $u^T(b - Ax) \geq 0$ and $x^T(c - A^T u) \leq 0$. This suggests that the fuzzy dual formulation of Section 5.2 should be modified suitably so that the crisp results follow as a special case. We present this modified construction in this section which is based on Bector and Chandra [4].

We already know that the crisp pair of primal-dual linear programming problems is

(LP)
$$\begin{aligned} \max \quad & c^T x \\ \text{subject to,} \quad & \\ & Ax \leq b, \\ & x \geq 0, \end{aligned}$$

(LD)
$$\begin{aligned} \min \quad & b^T w \\ \text{subject to,} \quad & \\ & A^T w \geq c, \\ & w \geq 0. \end{aligned}$$

To construct the fuzzy pair of such problems it seems natural to consider the fuzzy version problems (LP) and (LD) in the sense of Zimmermann [90]. Let us call them as $(\widetilde{LP})$ and $(\widetilde{LD})$. Then

$(\widetilde{LP})$ Find $x \in \mathbb{R}^n$ such that,
$$\begin{aligned} c^T x &\gtrsim Z_0, \\ Ax &\lesssim b, \\ x &\geq 0, \end{aligned}$$

and,

$(\widetilde{LD})$ Find $w \in \mathbb{R}^m$ such that,

$$
\begin{aligned}
b^T w &\lesssim W_0, \\
A^T w &\gtrsim c, \\
w &\geq 0.
\end{aligned}
$$

Here "$\gtrsim$" and $\lesssim$ are fuzzy version of symbols "$\geq$" and "$\leq$" respectively and have interpretation of "essentially greater than" and "essentially less than" in the sense of Zimmermann [90]. Also Z_0 and W_0 are the aspiration levels of the two objectives $c^T x$ and $b^T w$.

Now let p_0, p_i $(i = 1, 2, \ldots, m)$ be subjectively chosen constants of admissible violations associated with the objective function and constraints of (LP). Then we can choose Zimmermann's [90] type linear membership function $\mu_i(x)$, $(i = 0, 1, 2, \ldots, m)$ as follows

$$
\mu_o(x) = \begin{cases} 1 & , c^T x \geq Z_0, \\ 1 - \dfrac{Z_0 - c^T x}{p_0} & , Z_0 - p_0 \leq c^T x < Z_0, \\ 0 & , c^T x < Z_0 - p_0, \end{cases}
$$

and,

$$
\mu_i(x) = \begin{cases} 1 & , A_i x \leq b_i, \\ 1 - \dfrac{A_i x - b_i}{p_i} & , b_i < A_i x \leq b_i + p_i, \\ 0 & , A_i x > b_i + p_i. \end{cases}
$$

Hence using these membership functions μ_i and following Zimmermann [90], the crisp equivalent of the fuzzy linear programming problem $(\widetilde{LP})$ is

(CP) max λ

subject to,

$$
\begin{aligned}
(\lambda - 1) p_0 &\leq c^T x - Z_0, \\
(\lambda - 1) p_i &\leq b_i - A_i x, \quad (i = 1, 2, \ldots, m), \\
\lambda &\leq 1, \\
x, \lambda &\geq 0,
\end{aligned}
$$

where A_i is the ith row of matrix A and b_i is the ith component of b $(i = 1, 2, \ldots, m)$.

Similarly, let q_j $(j = 0, 1, 2, \ldots, n)$ be subjectively chosen constants of the admissible violations of the objective and the constraint functions of (LD). Then the crisp equivalent of the problem $(\widetilde{LD})$ is

(CD) min $-\eta$

subject to,

$$(\eta - 1)q_0 \leq W_0 - b^T w,$$
$$(\eta - 1)q_j \leq A_j^T w - c_j, \quad (j = 1, 2, \ldots, n),$$
$$\eta \leq 1,$$
$$x, \eta \geq 0,$$

where A_j^T denotes the jth row of the matrix A^T and c_j is the jth component of c $(j = 1, 2, \ldots, n)$. Since all p_i and q_j are positive, the problems (CP) and (CD) can be written as

(EP) max λ

subject to,

$$\lambda \leq 1 + \frac{(c^T x - Z_0)}{p_0},$$
$$\lambda \leq 1 + \frac{(b_i - A_i x)}{p_i}, \quad (i = 1, 2, \ldots, m),$$
$$\lambda \leq 1,$$
$$x, \lambda \geq 0,$$

and,

(ED) min $-\eta$

subject to,

$$\eta \leq 1 + \frac{(W_0 - b^T w)}{q_0},$$
$$\eta \leq 1 + \frac{(A_j^T w - c_j)}{q_j}, \quad (j = 1, 2, \ldots, n),$$
$$\eta \leq 1,$$
$$w, \eta \geq 0.$$

The pair (CP)-(CD) (or equivalently (EP)-(ED))is termed as the modified fuzzy pair of primal-dual linear programming problems. We shall now prove certain modified duality theorems for the pair (CP)-(CD) (or equivalently (EP)-(ED)) which take into consideration the fact that the problems $(\widetilde{LP})$ and $(\widetilde{LD})$ are fuzzified version of problems (LP) and (LD).

Theorem 5.3.1 (Modified weak duality theorem). *Let* (x, λ) *be feasible for* (CP) *and* (w, η) *be feasible for* (CD). *Then,*

$$(\lambda - 1)p^T w + (\eta - 1)q^T x \leq (b^T w - c^T x),$$

where $p^T = (p_1, p_2, \ldots, p_m)$ *and* $q^T = (q_1, q_2, \ldots, q_n)$.

Proof. Since $(x,\ \lambda)$ is feasible for (*CP*) and $(w,\ \eta)$ is feasible for (*CD*), we have

$$(\lambda - 1)p \le b - Ax,\ x \ge 0,$$

and,

$$(\eta - 1)q \le A^T w - c,\ w \ge 0,$$

which imply

$$(\lambda - 1)p^T w + x^T A^T w \le b^T w,$$

and,

$$(\eta - 1)q^T x - w^T Ax \le -c^T x.$$

But $w^T Ax = x^T A^T w$ and therefore the above inequalities yield

$$(\lambda - 1)p^T w + (\eta - 1)q^T x \le (b^T w - c^T x).$$

Remark 5.3.1. When $\lambda = 1$ and $\eta = 1$ (i.e. when the original problems are crisp) the inequality in Theorem 5.3.1 reduces to $c^T x \le b^T w$, which is the standard weak duality theorem in the crisp linear programming duality theory. Also for $0 \le \lambda < 1$ and $0 \le \eta < 1$, the situation remains fuzzy which, for given tolerance levels p and q, is quantified in the expression $(\lambda - 1)p^T w + (\eta - 1)q^T x$.

Remark 5.3.2. In addition to the above inequality of the modified weak duality theorem, using the constraints of (*CP*) and (*CD*), we can also show that $(\lambda - 1)p_0 + (\eta - 1)q_0 \le (c^T x - b^T w) + (W_0 - Z_0)$. This inequality relates the relative difference of aspiration levels Z_0 of $c^T x$, and W_0 of $b^T w$, respectively, in terms of their tolerance levels p_0 and q_0.

Corollary 5.3.1 *Let* $(\bar{x},\ \bar{\lambda})$ *be feasible for* (CP) *and* $(\bar{w},\ \bar{\eta})$ *be feasible for* (CD) *such that*

(i) $(\bar{\lambda} - 1)p^T \bar{w} + (\bar{\eta} - 1)q^T \bar{x} = (b^T \bar{w} - c^T \bar{x})$,
(ii) $(\bar{\lambda} - 1)p_0 + (\bar{\eta} - 1)q_0 = (c^T \bar{x} - b^T \bar{w}) + (W_0 - Z_0)$,
(iii) the aspiration levels Z_0 *and* W_0 *satisfy* $Z_0 - W_0 \le 0$.

Then $(\bar{x},\ \bar{\lambda})$ *is optimal to* (CP) *and* $(\bar{w},\ \bar{\eta})$ *be optimal to* (CD).

Proof. Let $(x,\ \lambda)$ be feasible for (*CP*) and $(w,\ \eta)$ be feasible for (*CD*). Then by Theorem 5.3.1

$$(\lambda - 1)p^T w + (\eta - 1)q^T x - (b^T w - c^T x) \le 0.$$

From (i) we are given that

$$(\bar{\lambda} - 1)p^T \bar{w} + (\bar{\eta} - 1)q^T \bar{x} = (b^T \bar{w} - c^T \bar{x}).$$

These relations imply that, for any feasible solution $(x,\ \lambda)$ of (*CP*) and for any feasible solution $(w,\ \eta)$ of (*CD*), we have

$(\lambda-1)p^T w+(\eta-1)q^T x-(b^T w-c^T x) \leq (\bar{\lambda}-1)p^T \bar{w}+(\bar{\eta}-1)q^T \bar{x}-(b^T \bar{w}-c^T \bar{x})$.
This implies, that $(\bar{x},\ \bar{\lambda},\ \bar{w},\ \bar{\eta})$ is optimal to the following problem whose maximum value is zero.

$$
(MP) \qquad \max \quad [(\lambda-1)p^T w + (\eta-1)q^T x - (b^T w - c^T x)]
$$

subject to,

$$
\begin{aligned}
&(\lambda-1)p_0 \leq c^T x - Z_0,\\
&(\eta-1)q_0 \leq W_0 - b^T w,\\
&(\lambda-1)p_i \leq b_i - A_i x, \quad (i=1,2,\ldots,m),\\
&(\eta-1)q_j \leq A_j^T w - c_j, \quad (j=1,2,\ldots,n),\\
&\lambda,\eta \leq 1,\\
&x,w \geq 0,\\
&\lambda,\eta \geq 0.
\end{aligned}
$$

Now, from the given condition (i),

$$
(\bar{\lambda}-1)p^T \bar{w} + (\bar{\eta}-1)q^T \bar{x} - (b^T \bar{w} - c^T \bar{x}) = 0.
$$

Also from the given condition (ii), we have

$$
(\bar{\lambda}-1)p_0 + (\bar{\eta}-1)q_0 - (W_0 - Z_0) - (c^T \bar{x} - b^T \bar{w}) = 0.
$$

Equating the above two expressions, we get

$$
(\bar{\lambda}-1)p^T \bar{w} + (\bar{\eta}-1)q^T \bar{x} + (\bar{\lambda}-1)p_o + (\bar{\eta}-1)q_o + (Z_o - W_o) = 0.
$$

But each term in the above sum is non-positive (because $\bar{\lambda},\bar{\eta} \leq 1$) and therefore, each of these should separately be equal to zero, i.e.

$$
\begin{aligned}
(\bar{\lambda}-1)p^T \bar{w} &= 0,\\
(\bar{\eta}-1)q^T \bar{x} &= 0,\\
(\bar{\lambda}-1)p_0 &= 0,\\
(\bar{\eta}-1)q_0 &= 0,\\
(Z_0 - W_0) &= 0.
\end{aligned}
$$

Since,

$$
(\lambda-1)p_0 \leq 0 \text{ and } (\eta-1)q_0 \leq 0,
$$

these imply,

$$
(\lambda-1)p_0 \leq (\bar{\lambda}-1)p_0,
$$

$$
(\eta-1)q_0 \leq (\bar{\eta}-1)q_0.
$$

But, $p_0 > 0$ and $q_0 > 0$, so cancelling p_0 and q_0 we see $\lambda \leq \bar{\lambda}$ and $-\eta \geq -\bar{\eta}$.

Remark 5.3.3. Since (CP) and (CD) are not duals in the conventional sense but are only the crisp equivalents of fuzzy pairs $(\widetilde{LP})$ and $(\widetilde{LD})$, there may not be any direct or converse duality theorem between them. Thus in the fuzzy scenario, we should not expect any strong duality theorem or equality of two objectives of (*EP*) and (*ED*) for respective optimal solutions $(\bar{x}, \bar{\lambda})$ and $(\bar{w}, \ \bar{\eta})$. However, by following the usual arguments of linear programming duality, one can prove that if (*CP*)(respectively (*CD*)) has an optimal solution then (*CD*)(respectively (*CP*)) will certainly have an feasible solution. Furthermore, if the feasible region of (*CP*)(respectively (*CD*)) is bounded, then (*CD*)(respectively (*CP*)) will have an optimal solution, but the value of the two objective functions will not be equal in general.

Now, the constraints of (*EP*) can be written as

$$\lambda p_i \leq p_i + (b_i - A_i x), \ (i = 1, 2 \ldots, m).$$

Therefore for any $w \in \mathbb{R}^m_+$, $w \neq 0$ from the above inequalities we have $\lambda p_i w_i \leq p_i w_i + (b_i - A_i x) w_i$, $(i = 1, 2 \ldots, m)$. Now summing over i, we have $\lambda \sum_{i=1}^{m} p_i w_i \leq \sum_{i=1}^{m} p_i w_i + \sum_{i=1}^{m} w_i (b_i - A_i x)$, which for $w \in \mathbb{R}^m_+$, $w \neq 0$ yields $\lambda \leq 1 + \dfrac{w^T(b - Ax)}{w^T p}$.

Therefore, the problem (*EP*) becomes

$$
\begin{aligned}
(EP1) \quad & \max \quad \lambda \\
& \text{subject to,} \\
& \lambda \leq 1 + \frac{(c^T x - Z_0)}{p_0}, \\
& \lambda \leq 1 + \frac{w^T(b - Ax)}{w^T p}, \ \text{(for any given } w \geq 0, \ w \neq 0), \\
& \lambda \leq 1, \\
& x \geq 0, \\
& \lambda \geq 0.
\end{aligned}
$$

Similarly, the problem (*ED*) can be written as

$$
\begin{aligned}
(ED1) \quad & \min \quad (-\eta) \\
& \text{subject to,} \\
& \eta \leq 1 + \frac{(W_0 - b^T w)}{q_0}, \\
& \eta \leq 1 + \frac{x^T(A^T w - c)}{q^T x} \ \text{(for any given } x \geq 0, x \neq 0), \\
& \eta \leq 1, \\
& x \geq 0, \\
& \eta \geq 0.
\end{aligned}
$$

It is interesting to see that problems (*EP*1) and (*ED*1) are very similar to problems (*FP*) and (*FD*) of Section 5.2 and are obtained if the membership functions $\mu^I(x)$, $\mu_x^I(w)$, $\mu^M(x)$, $\mu_w^M(x)$ are modified suitably.

Example 5.3.4. We now present a simple numerical example.

(*LP*)

$$\begin{aligned} \max \quad & 2x \\ \text{subject to,} \quad & \\ & x \leq 1, \\ & x \geq 0. \end{aligned}$$

and,

(*LD*)

$$\begin{aligned} \min \quad & w \\ \text{subject to,} \quad & \\ & w \geq 2, \\ & w \geq 0. \end{aligned}$$

Now taking $p_0 = 2$, $p_1 = 2$ and $Z_0 = 1$ for (*LP*), the corresponding problem (*CP*) becomes

(*CP*)

$$\begin{aligned} \max \quad & \lambda \\ \text{subject to,} \quad & \\ & 2\lambda - 2x \leq 1, \\ & 2\lambda + x \leq 3, \\ & \lambda \leq 1, \\ & x \geq 0, \\ & \lambda \geq 0. \end{aligned}$$

The optimal solution of (*CP*) is at $x^* = \dfrac{1}{2}$, $\lambda^* = 1$, and the optimal value of the (*CP*) is $\lambda^* = 1$.

Now taking $q_0 = 1$, $q_1 = 3$ and $W_0 = 1$ for (*LD*), the corresponding problem (*CD*) becomes

(*CD*)

$$\begin{aligned} \min \quad & (-\eta) \\ \text{subject to,} \quad & \\ & \eta + w \leq 2, \\ & 3\eta - w \leq 1, \\ & \eta \leq 1, \\ & \eta \geq 0, \\ & w \geq 0. \end{aligned}$$

The optimal solution of (CD) is at $w^* = \dfrac{5}{4}$, $\eta^* = \dfrac{3}{4}$, and the optimal value of the (CD) is $-\eta^* = -\frac{3}{4}$.

Also, the optimal value of (MP) is non-positive (≤ 0) for $x^* = \dfrac{1}{2}$, $\lambda^* = 1$, $w^* = \dfrac{5}{4}$, $\eta^* = \dfrac{3}{4}$, and it will remain so far all feasible solutions of (CP) and (CD).

5.4 Verdegay's dual for fuzzy linear programming

Let us recall from Chapter 4 the approach of Verdegay [74] to solve fuzzy linear programming problems. Specifically it has been shown there that a linear programming problem with fuzzy constrains could be solved by converting it into a linear programming with fuzzy objective function and vice-versa. This intuitively suggests that there is some duality between these two classes of fuzzy linear programming problems and in this section we attempt to bring out this fact more explicitly. For this, we shall continue with the notations of Section 4.4 and denote linear programming problems with fuzzy constraints by $(P1-FLP)$ and linear programming problems with fuzzy objective function by $(P2-FLP)$.

Theorem 5.4.1 *For a fuzzy linear programming problem of the type* (*P*1-*FLP*) *there always exists another one of the type* (*P*2-*FLP*) *having the same fuzzy solution.*

Proof. Consider a fuzzy linear programming of the type (*P*1-*FLP*) as follows

(*P*1-*FLP*1)
$$\begin{aligned} \max \quad & c^T x \\ \text{subject to,} \quad & \\ & Ax \lesssim b, \\ & x \geq 0, \end{aligned}$$

where $c \in \mathbb{R}^n$, $b \in \mathbb{R}^m$, $A \in \mathbb{R}^m \times \mathbb{R}^n$ and $x \in \mathbb{R}^n$. Also the fuzziness of the constraints is understood in the sense of following membership functions μ_i $(i = 1, 2, \dots, m)$, where

$$\mu_i(v) = \begin{cases} 1 & , v \leq b_i, \\ 1 + \dfrac{b_i - v}{p_i} & , b_i < v \leq b_i + p_i, \\ 0 & , v > b_i + p_i, \end{cases}$$

p_i $(i = 1, 2, \ldots, m)$ being the violation which the decision maker allows in the fulfillment of the ith constraint. Then as discussed in Chapter 4, by following Zimmermann's approach, the given fuzzy linear programming problem ($P1$-$FLP1$) can be solved by determining an optimal solution of linear parametric problem

$$(LP)_\alpha \qquad \max \quad c^T x$$

subject to,

$$\begin{aligned} Ax &\leq b + (1-\alpha)p, \\ x &\geq 0, \\ \alpha &\in [0,1], \end{aligned}$$

where $p = (p_1, p_2, \ldots, p_m)^T$ is the vector of tolerances.

But $(LP)_\alpha$ is a standard linear parametric problem whose dual is

$$(LD)_\alpha \qquad \min \quad \big(b + (1-\alpha)p\big)^T u$$

subject to,

$$\begin{aligned} A^T u &\geq c, \\ u &\geq 0, \\ \alpha &\in [0,1]. \end{aligned}$$

Let us next, denote the set $\{u \in \mathbb{R}^n : A^T u \geq c, \; u \geq 0\}$ by S. Then $(LD)_\alpha$ can equivalently be rewritten as

$$(ELD)_\alpha \qquad \min \quad a^T u$$

subject to,

$$\begin{aligned} a &= b + (1-\alpha)p, \\ u &\in S, \\ \alpha &\in [0,1]. \end{aligned}$$

If we now agree to take $\beta = (1-\alpha)$ and treat a as a variable (depending parametrically on β) then $(ELD)_\alpha$ becomes

$$(ELD)_\beta \qquad \min \quad a^T u$$

subject to,

$$\begin{aligned} a &\leq b + \beta p, \\ u &\in S, \\ \beta &\in [0,1]. \end{aligned}$$

Here it may be noted that in $(ELD)_\beta$, the first constraint can be taken as an inequality rather than equality because any optimal solution of

$(ELD)_\beta$ is also an optimal solution of $(ELD)_\alpha$. But for $i = 1, 2, \ldots, m$, $a_i \leq b_i + \beta p_i$ means $(b_i + p_i - a_i) \geq (1-\beta)p_i$ i.e. $1 + \dfrac{(b_i - a_i)}{p_i} \geq (1-\beta)$. Therefore from the definition of μ_i, we can express $(ELD)_\beta$ equivalently as

$(ELD)_\beta$ min $a^T u$

subject to,

$$\begin{aligned} \mu_i(a_i) &\geq 1 - \beta, \; (i = 1, 2, \ldots, m), \\ u &\in S, \\ \beta &\in [0, 1], \end{aligned}$$

which is the (crisp) linear parametric problem for the fuzzy linear programming problem of the type (*P*2-*FLP*), say (*P*2-*FLD*1), where,

(*P*2-*FLD*1) $\widetilde{min}$ $a^T u$

subject to,

$$\begin{aligned} A^T u &\geq c, \\ u &\geq 0. \end{aligned}$$

Since, by the crisp linear programming duality problem $(LP)_\alpha$ and $(LD)_\alpha$ have the same parametric optimal values, the problems (*P*1-*FLP*1) and (*P*2-*FLD*1) have the same fuzzy solution with $\beta = (1 - \alpha)$.

Remark 5.4.1. The converse of the Theorem 5.4.1 also holds, i.e. for a fuzzy linear programming problem of the type (*P*2-*FLP*), there always exists another of the type (*P*1-*FLP*) having the same fuzzy solutions. The proof follows on the lines of the proof of Theorem 5.4.1. Therefore the pair (*P*1-*FLP*1) and (*P*2-*FLD*1) is called the Verdegay's primal-dual pair of fuzzy linear programming problems.

Although the duality result between (*P*1-*FLP*1) and (*P*2-*FLD*1) has been proved by taking the membership functions μ_i $(i = 1, 2, \ldots, m)$ as linear, it can be shown that the proof does not depend on this assumption about μ_i. This we state in the form of the following theorem.

Theorem 5.4.2 *For a fuzzy linear programming problem of the type* (P1-FLP) *(respectively* (P2-FLP)*) with continuous and strictly monotone membership functions for the constraints (respectively objective function) there always exists another fuzzy linear programming problem of the type* (P2-FLP) *(respectively* (P1-FLP)*), called the dual of the given problem, such that both problems have the same fuzzy solution.*

Proof. let $\mu_i : \mathbb{R} \to [0,1]$, $(i = 1,2,\ldots,m)$, be continuous and strictly increasing function of the given fuzzy linear programming problem,

$$\begin{aligned} &\max \quad c^T x \\ &\text{subject to,} \\ &Ax \lesssim b, \\ &x \geq 0. \end{aligned}$$

We shall find its fuzzy solution from every α-cut of the fuzzy constraints set $\mu(Ax,b) \geq \alpha$, $\alpha \in [0,1]$. But then by the assumptions on μ, we have $\mu(Ax,b) \geq \alpha \Leftrightarrow Ax \leq \phi(\alpha) = \mu^{-1}(\alpha)$, and then the proof follows exactly the same way as in Theorem 5.4.1.

Example 5.4.2. (Verdegay [74]). Let us verify Theorem 5.4.1 and Theorem 5.4.2 for the following problem:

$$\begin{aligned} &\max \quad z = x_1 + x_2 \\ &\text{subject to,} \\ &4x_1 - x_2 \lesssim 10, \\ &x_1 + 2x_2 \lesssim 15, \\ &5x_1 + 2x_2 \lesssim 20, \\ &x_1, x_2 \geq 0. \end{aligned}$$

Let us take the membership functions as

$$\begin{aligned} \mu_1(4x_1 - x_2, 10) &= \frac{(15 - 4x_1 + x_2)^2}{25}, \quad 10 \leq 4x_1 - x_2 \leq 15, \\ \mu_2(x_1 + 2x_2, 15) &= \frac{(23 - x_1 - 2x_2)^2}{64}, \quad 15 \leq x_1 + 2x_2 \leq 23, \\ \mu_3(5x_1 + 2x_2, 20) &= \frac{(30 - 5x_1 - 2x_2)^2}{100}, \quad 20 \leq 5x_1 + 2x_2 \leq 30, \end{aligned}$$

where μ_i $(i = 1,2,3)$ takes value 0 and 1 outside these intervals, as usual. On solving the given fuzzy linear programming problem (which is of the type (*P*1-*FLP*1)) by employing the above membership functions, we get

$$x_1^* = \frac{7 - 2\sqrt{\alpha}}{4}, \ x_2^* = \frac{85 - 30\sqrt{\alpha}}{8}, \ \alpha \in [0,1].$$

Thus, $z^* = x_1^* + x_2^* = \dfrac{99 - 34\sqrt{\alpha}}{8} \in \left[\dfrac{65}{8}, \dfrac{99}{8}\right]$, and the fuzzy solution becomes the fuzzy set $\left\{ \left(x,\ \mu(x)\right) :\ x \in \left[\dfrac{65}{8}, \dfrac{99}{8}\right],\ \mu(x) = \left(\dfrac{99 - 8x}{34}\right)^2 \right\}$.

If we now write the dual of the given problem to get (*P2-FLD1*), we get

$$\begin{aligned} \min \quad & w = (15-5\sqrt{\alpha})u_1+(23-8\sqrt{\alpha})u_2+(30-10\sqrt{\alpha})u_3) \\ \text{subject to,} \quad & \\ & 4u_1 + u_2 + 5u_3 \geq 1, \\ & -u_1 + 2u_2 + 2u_3 \geq 1, \\ & \alpha \in [0,1], \\ & u_1,\ u_2,\ u_3 \geq 0. \end{aligned}$$

On solving the above parametric problem we get $u_1^* = 0$, $u_2^* = \frac{3}{8}$, $u_3^* = \frac{1}{8}$ and $w^* = (15 - 5\sqrt{\alpha})u_1^* + (23 - 8\sqrt{\alpha})u_2^* + (30 - 10\sqrt{\alpha})u_3^* = (99 - 34\sqrt{\alpha}) \in \left[\frac{65}{8}, \frac{99}{8}\right]$, $\alpha \in [0,1]$ and the corresponding fuzzy solution is the fuzzy set $\left\{ \left(x,\ \mu(x)\right) :\ x \in \left[\frac{65}{8}, \frac{99}{8}\right],\ \mu(x) = \left(\frac{99 - 8x}{34}\right)^2 \right\}$ which is the same as obtained earlier for the given primal problem.

5.5 Duality for quadratic programming under fuzzy environment

In this section, we extend the approach of Section 5.3 to study duality for quadratic programming under fuzzy environment . In the crisp case, quadratic programming duality is well established, (e.g. Mangasarian [53]) for the primal-dual pair $(QP) - (QD)$ where

(*QP*)

$$\begin{aligned} \max \quad & c^T x - \tfrac{1}{2}x^T H x \\ \text{subject to,} \quad & \\ & Ax \leq b, \\ & x \geq 0, \end{aligned}$$

and,

(*QD*)

$$\begin{aligned} \min \quad & b^T u + \tfrac{1}{2}w^T H w \\ \text{subject to,} \quad & \\ & A^T u + Hw \geq c, \\ & u \geq 0, \\ & w \geq 0. \end{aligned}$$

Here $u \in \mathbb{R}^m$, $w \in \mathbb{R}^n$, $H \in \mathbb{R}^n \times \mathbb{R}^n$ is a positive semidefinite matrix. The other symbols c, b, x and A are as described in Section 5.3.

Now taking aspiration levels for the objective functions of (QP) and (QD) as Z_0 and W_0 respectively, we consider the fuzzy version of these problems $\big($say $(\widetilde{QP})$ and $(\widetilde{QD})\big)$ as follows:

$(\widetilde{QP})$ Find $x \in \mathbb{R}^n$ such that

$$\begin{aligned} c^T x - \tfrac{1}{2}x^T H x &\gtrsim Z_o \\ Ax &\lesssim b, \\ x &\geq 0, \end{aligned}$$

and,

$(\widetilde{QD})$ Find $(u, w) \in \mathbb{R}^m \times \mathbb{R}^n$ such that,

$$\begin{aligned} b^T u + \tfrac{1}{2}w^T H w &\lesssim W_o \\ A^T u + Hw &\gtrsim c, \\ u &\geq 0, \\ w &\geq 0. \end{aligned}$$

Here as in Section 5.3, "$\gtrsim$" and "$\lesssim$" are fuzzy version of symbols "essentially greater than" and "essentially less than" as explained there.

Now let $f(x) = c^T x - \frac{1}{2}x^T H x$ and $F(u, w) = b^T u + \frac{1}{2}w^T H w$. Also let $p_i > 0$ $(i = 0, 1, \ldots, m)$ be subjectively chosen constants of admissible violations such that p_0 is associated with the objective function and p_i $(i = 1, 2, \ldots, m)$ is associated with the ith constraint of (QP). Then, as in Section 5.3, using the membership function which increases linearly over the "tolerance interval" p_i $(i = 0, 1, 2, \ldots, m)$, the crisp formulation of the fuzzy quadratic programming problem $(\widetilde{QP})$ is as follows (Zimmermann [90]).

(MP) max λ
subject to,

$$\begin{aligned} (\lambda - 1)p_0 &\leq f(x) - Z_o, \\ (\lambda - 1)p_i &\leq b_i - A_i x, \quad (i = 1, 2, \ldots, m), \\ \lambda &\leq 1, \\ x, \lambda &\geq 0, \end{aligned}$$

where A_i $(i = 1, 2, \ldots, m)$ denotes the ith row of the matrix A and b_i $(i = 1, 2, \ldots, m)$ denotes the ith component of b.

Similarly, let $q_j > 0$ $(j = 0, 1, \ldots, n)$ be subjectively chosen constants of admissible violations of the objective and the constraints of (QD). Then the crisp formulation of the problem $(\widetilde{QD})$ is as follows

(*MD*) min $(-\eta)$
subject to,

$$\begin{aligned}
(\eta-1)q_0 &\le W_o - F(u,w),\\
(\eta-1)q_j &\le H_j^T w + A_j^T u - c_j, \ (j=1,2,\ldots,n),\\
\eta &\le 1,\\
x, w, \eta &\geqq 0,
\end{aligned}$$

where H_j^T and A_j^T $(j=1,2,\ldots,n)$ denote the jth row of the matrix H^T and A^T respectively, and c_j $(j=1,2,\ldots,n)$ denotes the jth component of c. Since the tolerances p_i $(i=0,1,\ldots,m)$ and q_j $(j=0,1,\ldots,n)$ are positive, problems (*MP*) and (*MD*) can be rewritten as
(*MP*1) max λ
subject to,

$$\begin{aligned}
\lambda &\le 1 + \frac{f(x)-Z_0}{p_0},\\
\lambda &\le 1 + \frac{b_i - A_i x}{p_i}, \quad (i=1,2,\ldots,m),\\
\lambda &\le 1,\\
x, \lambda &\ge 0,
\end{aligned}$$

and
(*MD*1) min $(-\eta)$
subject to,

$$\begin{aligned}
\eta &\le 1 + \frac{W_0 - F(u,w)}{q_0},\\
\eta &\le 1 + \frac{H_j^T w + A_j^T u - c_j}{q_j}, \ (j=1,2,\ldots,n),\\
\eta &\le 1,\\
w, \eta &\ge 0.
\end{aligned}$$

The pair (*MP*)-(*MD*) (or equivalently (*MP*1)-(*MD*1)) are termed as the fuzzy pair of primal-dual quadratic programming problems.

Now similar to results on fuzzy linear programming (Theorem 5.3.1), we shall prove certain duality theorems for the pair (*MP*)-(*MD*) (or (*MP*1)-(*MD*1)) which take into consideration that the problems $(\widetilde{QP})$ and $(\widetilde{QD})$ are the fuzzified version of problems (*QP*) and (*QD*).

Theorem 5.5.1 (Weak duality). *Let* $(x,\ \lambda)$ *be feasible for* (*MP*1) *and* $(u,\ w,\ \eta)$ *be feasible for* (*MD*1)*. Then,* *(i)* $(\lambda - 1)\sum_{i=1}^{m} p_i u_i + (\eta - 1)\sum_{j=1}^{n} q_j x_j \leq F(u, w) - f(x)$.

(ii) $(\lambda - 1)p_0 + (\eta - 1)q_0 \leq (f(x) - F(u, w)) + (W_0 - Z_0)$.

Proof. The proof is similar to that of Theorem 5.3.1.

We can also prove other duality related results for (*MP*1)-(*MD*1) on the lines of those proved in Section 5.3 for the case of fuzzy linear programming.

Remark 5.5.1. The weak duality theorem takes into consideration the fuzzy scenario. This can be checked from the fact for $\lambda = 1$ and $\eta = 1$ (i.e. when the original problems are crisp) the inequality reduces to $f(x) \leq F(u, w)$, which is the standard weak duality theorem in a crisp quadratic program. For $0 \leq \lambda < 1$ and $0 \leq \eta < 1$, the situation remains fuzzy which, for the given tolerance levels p_i $(i = 1, 2, \ldots m)$, and q_j $(j = 1, 2, \ldots n)$, is quantified in the expression $(\lambda - 1)\sum_{i=1}^{m} p_i w_i + (\eta - 1)\sum_{j=1}^{n} q_j x_j$.

Remark 5.5.2. Similar to fuzzy linear programming case there may not be any direct or converse duality theorem between the pair (*MP*1)-(*MD*1) and the two objective function values may not be equal.

Remark 5.5.3. For the case when $H = 0$, the problems (*QP*) and (*QD*) become the linear programming problems (*LP*)-(*LD*). In this situation, results of this section reduce to those of Section 5.3.

Example 5.5.4. In this section we present a simple numerical example to illustrate the construction of the fuzzy primal-dual pair and to verify the modified weak duality theorem. For this, we consider the problem (*QP*) and its dual (*QD*) as follows.

(*QP*) max $f(x) = 2x - \frac{1}{2}x^2$
subject to,

$$x \leq 1,$$
$$x \geq 0,$$

and,

(*QD*) min $F(u, w) = u + \frac{1}{2}w^2$
subject to,

$$u + w \geq 2,$$
$$u, w \geq 0.$$

Now taking $p_0 = 2$, $p_1 = 2$ and $Z_0 = 1$ for (QP), the corresponding problem $(MP1)$ becomes

$$(MP1) \qquad \max \quad \lambda$$
$$\text{subject to,}$$
$$2\lambda - 2x + \tfrac{1}{2}x^2 \leq 1,$$
$$2\lambda + x \leq 3,$$
$$\lambda \leq 1,$$
$$x, \lambda \geq 0.$$

The optimal solution of $(MP1)$ is at $x^* = 1$, $\lambda^* = 1$, and the optimal value of the $(MP1)$ is $\lambda^* = 1$.

Now taking $q_0 = 1$, $q_1 = 3$ and $W_0 = 1$ for (QD), the corresponding problem $(MD1)$ becomes

$$(MD1) \qquad \min \quad (-\eta)$$
$$\text{subject to,}$$
$$\eta + u + \tfrac{1}{2}w^2 \leq 2,$$
$$3\eta - u - w \leq 1,$$
$$\eta \leq 1,$$
$$u, \eta, w \geq 0.$$

The optimal solution of $(MD1)$ is at $\eta^* = 0.875$, $u^* = 0.625$, $w^* = 1$ and the optimal value of the $(MD1)$ is $-\eta^* = -0.875$.

5.6 Conclusions

In this chapter we have presented certain very basic models of linear and quadratic programming duality under fuzzy environment. The discussion on fuzzy linear programming duality will be continued in Chapter 7 as well where linear programming problems with fuzzy parameters will be considered and appropriate duality results will be proved. We shall come back to this topic again in Chapter 10 where the results will be established in a more general setting of fuzzy (valued) relations etc. In the crisp scenario, the duality results have been applied to study sensitivity analysis/post optimality analysis in linear programming, e.g. Bazaraa et al. [2]. Such a study is also possible for linear programming problems under fuzzy environment, for which an appropriate reference is Hamacher et al. [22].

6

Matrix games with fuzzy goals

6.1 Introduction

Let us recall our discussion on (crisp) two-person zero-sum matrix game theory from Chapter 1 and take note of one of the most celebrated and useful result which asserts that every two person zero sum matrix game is equivalent to two linear programming problems which are dual to each other. Thus, solving such a game amounts to solving any one of these two mutually dual linear programming problems and obtaining the solution of the other by using linear programming duality theory.

Although various attempts have been made in the literature to study two person zero sum fuzzy matrix games (for example, Campos [10], Nishizaki and Sakawa [61] and Sakawa and Nishizaki [68]) but they do not take into consideration the fuzzy linear programming duality aspects. In this context it may be noted that the fuzzy linear programming duality results are available in the literature and some of these have already been discussed in Chapter 5 but unlike their crisp counter parts, they have not been used for the study of fuzzy matrix game theory until very recently.

Now, similar to fuzzy linear programming problems, fuzziness in matrix games can also appear in so many ways but two cases of fuzziness seem to be very natural. These being the one in which players have fuzzy goals and the other in which the elements of the pay-off matrix are given by fuzzy numbers. These two classes of fuzzy matrix games are referred as *matrix games with fuzzy goals* and *matrix games with fuzzy pay-offs* respectively. However in our presentation, the term *fuzzy matrix game* will often be used in a general sense and the context will specify if the goals are fuzzy or the pay-offs are fuzzy or goals and pay-offs, both are fuzzy.

This chapter presents two models for conceptualizing a two person zero sum game with fuzzy goals; one due to Nishizaki and Sakawa [61] and the other due to Bector, Chandra and Vijay [6]. Although these two models are similar in their approaches, the Nishizaki and Sakawa's model does not use fuzzy linear programming duality and is also somewhat restricted because it needs certain modifications of the matrix game to capture the crisp scenario as well. The other model [6] is more general and it shows that as in the crisp environment, in the fuzzy environment as well, there is a complete equivalence between a matrix game with fuzzy goals and a primal-dual pair of fuzzy linear programming problems.

This chapter is divided into three main sections, namely, *two person zero sum matrix game with fuzzy goals: a generalized model, two person zero sum matrix game with fuzzy goals: Nishizaki and Sakawa's model,* and *special cases.*

6.2 Matrix game with fuzzy goals: a generalized model

The required literature regarding two person zero sum (crisp) matrix game theory and duality in linear programming under fuzzy environment in the sense of Bector and Chandra [4] has already been discussed in Chapter 1 and Chapter 5 respectively. Therefore, let S^m, S^n and A be as introduced in Section 1.3. Let v_0, w_0 be scalers representing the aspiration levels of Player I and Player II respectively. Then a *two person zero-sum matrix game with fuzzy goals*, denoted by FG, is defined as

$$FG = (S^m,\ S^n,\ A,\ v_0,\ \gtrsim,\ ;w_0,\ \lesssim),$$

where as explained earlier, '$\gtrsim$' and '$\lesssim$' are fuzzified versions of $\geq$ and $\leq$ respectively. Therefore the game FG gets fixed only when the specific choices of membership functions are made to define '$\gtrsim$' and '$\lesssim$' in the sense of Zimmermann [91].

Let t be a real variable and $a \in R$. Let $p > 0$, then the fuzzy set F defining the fuzzy statement $t \gtrsim_p a$, to be read as "t essentially greater than or equal to a with tolerance error p ", is to be understood in terms of the following membership function

$$\mu_F(t) = \begin{cases} 1 & , t \geq a \\ 1 - \dfrac{a-t}{p} & , (a-p) \leq t < a \\ 0 & , t < (a-p). \end{cases}$$

Lemma 6.2.1, as stated below, follows from the definition of $\mu_F(t)$.

Lemma 6.2.1. *Let* $t_1 \gtrsim_p a$, $t_2 \gtrsim_p a$, $\alpha \geq 0$, $\beta \geq 0$ *and* $\alpha + \beta = 1$. *Then* $\alpha t_1 + \beta t_2 \gtrsim_p a$.

Proof. Relations $t_1 \gtrsim_p a$, and $t_2 \gtrsim_p a$, can be written as

$$t_1 \geq a - (\lambda - 1)p$$
$$t_2 \geq a - (\lambda - 1)p.$$

Then

$$\alpha t_1 + \beta t_2 \geq (\alpha + \beta)a - (\lambda - 1)(\alpha + \beta)p,$$

which gives

$$\alpha t_1 + \beta t_2 \gtrsim_p a.$$

In view of the above discussion we include tolerances p_0 and q_0 for Player I and Player II respectively in our definition of the fuzzy game *FG* and therefore take *FG* as

$$FG = (S^m,\ S^n,\ A,\ v_0,\ \gtrsim,\ p_0,\ w_0,\ \lesssim,\ q_0).$$

Now we define the meaning of the "*solution*" of the fuzzy matrix game *FG*.

Definition 6.2.1 (Solution of the fuzzy matrix game *FG***).** *A point* $(\bar{x}, \bar{y})$ *is called a solution to the fuzzy matrix game FG if*

$$(\bar{x})^T A y \gtrsim_{p_0} v_0, \text{ for all } y \in S^n,$$

and,

$$x^T A \bar{y} \lesssim_{q_0} w_0, \text{ for all } x \in S^m.$$

Since S^m and S^n are convex polytopes, for the choice of membership functions of type $\mu_F(t)$, Lemma 6.2.1 guarantees that in the Definition 6.2.1 it is sufficient to consider only the extreme points (i.e. pure strategies) of S^m and S^n. This observation leads to the following two fuzzy linear programming problems, (*FP*-1) and (*FP*-2), for Player I and Player II respectively

(*FP*-1) Find $x \in R^m$ such that,

$$\sum_{i=1}^{m} a_{ij}x_i \gtrsim_{p_0} v_0, \ (j = 1, 2, \ldots, n),$$
$$\sum_{i=1}^{m} x_i = 1,$$
$$x \geq 0,$$

and

(*FP*-2) Find $y \in R^n$ such that,

$$\sum_{j=1}^{n} a_{ij}y_j \lesssim_{q_0} w_0, \ (i = 1, 2, \ldots, m),$$
$$\sum_{j=1}^{n} y_j = 1,$$
$$y \geq 0.$$

Now noting that for $j = 1, 2, \ldots, n$, A_j denotes the jth column of A, the jth constraint of (*FP*-1) can be written as $A_j^T x \gtrsim_{p_0} v_0$. Similarly the ith constraint of (*FD*-2) can be written as $A_i y \lesssim_{q_0} w_0$ for $i = 1, 2, \ldots, m$. Therefore as per the requirement for the use of Lemma 6.2.1, we define the membership function $\mu_j(A_j^T x)$, $(j = 1, 2, \ldots, n)$ which gives the degree to which x satisfies the fuzzy constraint $A_j^T x \gtrsim_{p_0} v_0$, as follows:

$$\mu_j(A_j^T x) = \begin{cases} 1 & , A_j^T x \geq v_0 \\ 1 - \dfrac{v_0 - A_j^T x}{p_0} & , (v_0 - p_0) \leq A_j^T x < v_0 \\ 0 & , A_j^T x < (v_0 - p_0). \end{cases}$$

This choice of $\mu_j(A_j^T x)$ gives the crisp formulation of the fuzzy linear programming (*FP*-1) as:

(*FLP*) max λ

subject to,

$$\lambda \leq 1 - \frac{v_0 - A_j^T x}{p_0}, \quad (j = 1, 2, \ldots, n),$$
$$e^T x = 1,$$
$$\lambda \leq 1,$$
$$x, \lambda \geq 0.$$

Similarly the membership function $\nu_i(A_i y)$ $(i = 1, 2, ..., m)$, which defines the degree to which y satisfies the constraint $A_i y \lesssim_{q_0} w_0$, is taken as:

$$\nu_i(A_i y) = \begin{cases} 1 & , A_i y \leq w_0, \\ 1 - \dfrac{A_i y - w_0}{q_0} & , w_0 < A_i y \leq w_0 + q_0, \\ 0 & , A_i y > w_0 + q_0, \end{cases}$$

and this gives the crisp formulation of (*FD*-2) as:

$$\begin{array}{lll} (FLD) & \max & \eta \\ & \text{subject to,} & \\ & \eta \leq 1 - \dfrac{(A_i y - w_0)}{q_0}, & (i = 1, 2, \ldots, m), \\ & e^T y = 1, & \\ & \eta \leq 1, & \\ & y, \eta \geq 0. & \end{array}$$

From the above discussion we now observe that for solving the fuzzy matrix game *FG* we have to solve the crisp linear programming problems (*FLP*) and (*FLD*) for Player I and Player II respectively. Also if (x^*, λ^*) is an optimal solution of (*FLP*) then x^* is an optimal strategy for Player I and λ^* is the degree to which the aspiration level v_0 of Player *I* can be met by choosing to play the strategy x^*. Similar interpretations can also be given to an optimal solution (y^*, η^*) of (*FLD*). This leads to the following theorem.

Theorem 6.2.1 *The pair* (*FLP*) *and* (*FLD*) *constitutes a fuzzy primal-dual pair in the sense of Theorem 5.3.1.*

Proof. The proof follows by noting that (*FLD*) can be written as,

$$\begin{array}{lll} (FLP) & -\min & (-\eta) \\ & \text{subject to,} & \\ & \eta \leq 1 + \dfrac{(w_0 - A_i y)}{q_0}, & (i = 1, 2, \ldots, m), \\ & e^T y = 1, & \\ & \eta \leq 1, & \\ & y \geq 0, & \\ & \eta \geq 0, & \end{array}$$

and the comparing (*FLP*)-(*FLD*) with (*CP*)-(*CD*) as described in Section 5.3.

All the results discussed in this section can now be summarized in the form of Theorem 6.2.2 given below.

Theorem 6.2.2 *The fuzzy matrix game FG described by FG* = (S^m, S^n, A, v_0, $\gtrsim$, p_0, w_0, $\lesssim$, q_0) *is equivalent to two crisp linear programming problems* (*FLP*) *and* (*FLD*) *which constitute a primal-dual pair in the sense of duality for linear programming in a fuzzy environment [4].*

Remark 6.2.2. It is important to note that the crisp problems (*FLP*) and (*FLD*) do not constitute a primal-dual pair in the conventional sense of duality in linear programming but are dual in "fuzzy" sense as explained above.

Remark 6.2.3. If both players have the same aspiration levels, i.e. $v_0 = w_0$ and in the optimal solutions of (*FLP*) and (*FLD*) $\lambda^* = \eta^* = 1$, then the fuzzy game reduces to the crisp two person zero sum game G. Thus for $v_0 = w_0$, $\lambda^* = \eta^* = 1$, *FG* reduces to G; the pair (*FLP*)-(*FLD*) reduces to the pair (*LP*)-(*LD*); and as it should be, Theorem 6.2.2 reduces to Theorem 1.4.1.

6.3 Matrix game with fuzzy goals: Nishizaki and Sakawa model

This model is based on the maxmin and minmax principles of the crisp matrix game theory. In the crisp matrix game theory, for any pair $(x,\ y) \in S^m \times S^n$, $x^T Ay$ is the *expected pay-off* and the solution of the game is defined via the maxmin and minmax principles as described in Chapter 1.

Now we define the meaning of a fuzzy goal and try to explain how the players will play the game in a fuzzy environment.

Definition 6.3.1 (Fuzzy goal). *Let* $D = \{x^T Ay : (x,y) \in S^m \times S^n\} \subseteq \mathbb{R}$. *Then a fuzzy goal for Player I is a fuzzy set on D characterized by a membership function* $\mu_1 : D \longrightarrow [0,1]$. *Similarly, a fuzzy goal for Player II is also a fuzzy set on D, characterized by a membership function* $\mu_2 : D \longrightarrow [0,1]$.

A membership function value for a fuzzy goal can be interpreted as the degree of attainment of the fuzzy goal for the pay-off. Therefore when a player has two different payoffs, he prefers the payoff possessing the higher membership function value in comparison to the other. It means that Player I aims to maximize the degree of attainment for his fuzzy goal.

We assume that Player I supposes that Player II will choose a strategy y so as to minimize Player I's degree of attainment of the fuzzy goal. Assuming that Player I chooses $x \in S^m$, the least degree of attainment of his goal will be $v(x) = \min_{y \in S^n} \mu_1(x^T Ay)$. Hence Player I will choose a strategy so as to maximize his degree of attainment of the fuzzy goal $v(x)$. In short, we assume that Player I behaves according to the maxmin principle in terms of degree of attainment of his fuzzy goal. Similar arguments hold for Player II as well.

Definition 6.3.2 (Maxmin value). *The maxmin value with respect to a degree of attainment of the fuzzy goal for Player I is defined as*

$$\max_{x \in S^m} \min_{y \in S^n} \mu_1(x^T Ay).$$

Similarly the maxmin value with respect to a degree of attainment of the fuzzy goal for Player II is defined as

$$\max_{y \in S^n} \min_{x \in S^m} \mu_2(x^T Ay).$$

Here it may be noted that in Nishizaki and Sakawa [61], for Player II, the minmax value is suggested but later it has been corrected to maxmin value for Player II as it should be.

Thus Player I (respectively Player II) wishes to determine $x^* \in S^m$ (respectively $y^* \in S^n$) such that the maxmin value with respect to the degree of attainment of the fuzzy goal for Player I (respectively Player II) is attained. This is another analogue of Definition 6.2.1 for the fuzzy matrix game FG.

We now analyze the optimization problems for Player I and Player II so as to obtain a solution of the given fuzzy game. For this we assume that membership functions of fuzzy goals $\mu_1(x^T Ay)$ and $\mu_2(x^T Ay)$ for Player I and Player II respectively are linear.

Optimization problem for Player I

Consider the membership function of the fuzzy goal $\mu_1(x^T Ay)$ for Player I described below,

$$\mu_1(x^TAy) = \begin{cases} 0 & , x^TAy \le \underline{a}, \\ 1 - \dfrac{\bar{a} - x^TAy}{\bar{a} - \underline{a}} & , \underline{a} < x^TAy \le \bar{a}, \\ 1 & , \bar{a} < x^TAy, \end{cases}$$

where $\bar{a}$ and $\underline{a}$ are the pay-offs giving the best and worst degree of satisfaction to Player I.

It is suggested in Nishizaki and Sakawa [61], that parameters $\bar{a}$ and $\underline{a}$ can be taken as

$$\underline{a} = \min_x \min_y \; x^TAy = \min_i \min_j \; a_{ij}$$

$$\bar{a} = \max_x \max_y \; x^TAy = \max_i \max_j \; a_{ij}$$

Theorem 6.3.1 *For the two person zero-sum fuzzy game FG, let the membership function μ_1 for Player I be linear as described above. Then Player I's maxmin solution with respect to the degree of attainment of the fuzzy goal is obtained by solving the following linear programming problem*

$$\begin{aligned} &\textit{max} \qquad \lambda \\ &\textit{subject to,} \\ &\sum_{i=1}^{m} \frac{a_{ij}}{\bar{a} - \underline{a}} x_i - \frac{\underline{a}}{\bar{a} - \underline{a}} \ge \lambda, \; (j = 1, 2, \dots, n), \\ &e^T x = 1, \\ &x \ge 0. \end{aligned}$$

Proof. The maxmin problem for Player I is

$$\max_{x \in S^m} \min_{y \in S^n} \; \mu_1(x^TAy),$$

which can be transformed into

$$\begin{aligned} \max_{x \in S^m} \min_{y \in S^n} \left(1 - \frac{\bar{a} - x^TAy}{\bar{a} - \underline{a}}\right) &= \max_{x \in S^m} \min_{y \in S^n} \left(\sum_{i=1}^{m} \sum_{j=1}^{n} \widehat{a_{ij}} x_i y_j + c\right) \\ &= \max_{x \in S^m} \; \min_{y \in S^n} \sum_{j=1}^{n} \left(\sum_{i=1}^{m} \widehat{a_{ij}} x_i + c\right) y_j \\ &= \max_{x \in S^m} \; \min_{j \in J} \left(\sum_{i=1}^{m} \widehat{a_{ij}} x_i + c\right), \end{aligned}$$

where $\widehat{a_{ij}} = \dfrac{a_{ij}}{\bar{a} - \underline{a}}$ and $c = -\dfrac{\underline{a}}{\bar{a} - \underline{a}}$.

Therefore taking $\min_j \left(\sum_{i=1}^{m} \widehat{a_{ij}} x_i + c \right) = \lambda$, the maxmin problem for Player I reduces to the desired linear programming problem.

Optimization Problem for Player II

Next, we consider Player II's maxmin solution with respect to the degree of attainment of his fuzzy goal. The membership function for the fuzzy goal $\mu_2(x^T Ay)$ can be represented as

$$\mu_2(x^T Ay) = \begin{cases} 1 & , x^T Ay \leq \underline{a}, \\ 1 - \dfrac{x^T Ay - \underline{a}}{\bar{a} - \underline{a}} & , \underline{a} < x^T Ay \leq \bar{a}, \\ 0 & , \bar{a} < x^T Ay. \end{cases}$$

Theorem 6.3.2 *For the two person zero-sum fuzzy game FG, let the membership function μ_2 for Player II be linear as described above. Then Player II's maxmin solution with respect to the degree of attainment of the fuzzy goal is obtained by solving the following linear programming problem:*

$$\begin{aligned} &\textit{min} \quad \lambda \\ &\textit{subject to,} \\ &\quad \sum_{j=1}^{n} \frac{a_{ij}}{\bar{a} - \underline{a}} y_j - \frac{\underline{a}}{\bar{a} - \underline{a}} \leq \lambda, \ (i = 1, 2, \ldots, m), \\ &\quad e^T y = 1, \\ &\quad y \geq 0. \end{aligned}$$

Proof. The maxmin problem for Player II is

$$\max_{y \in S^n} \min_{x \in S^m} \mu_2(x^T Ay),$$

which becomes,

$$\max_{y\in S^n}\min_{x\in S^m}\left(1-\frac{x^TAy-\underline{a}}{\bar{a}-\underline{a}}\right)=\max_{y\in S^n}\min_{x\in S^m}\left(-\sum_{i=1}^{m}\sum_{j=1}^{n}\widehat{a_{ij}}x_iy_j+1-c\right)$$
$$=\max_{y\in S^n}\min_{i\in I}\left(-\sum_{j=1}^{n}\widehat{a_{ij}}y_j+1-c\right),$$

where as before $\widehat{a_{ij}}=\dfrac{a_{ij}}{\bar{a}-\underline{a}}$ and $c=-\dfrac{\underline{a}}{\bar{a}-\underline{a}}$.

The strategy y^* satisfying the above is obtained by solving the following linear programming problem:

$$
\begin{aligned}
&\max \quad \eta\\
&\text{subject to,}\\
&\sum_{j=1}^{n}\frac{a_{ij}}{(\bar{a}-\underline{a})}y_j-\frac{\underline{a}}{(\bar{a}-\underline{a})}\leq 1-\eta, \quad (i=1,2,...,m),\\
&e^Ty=1,\\
&y\geq 0,
\end{aligned}
$$

which by taking $(1-\eta)=\lambda$, is equivalent to the linear programming problem given in the statement of the theorem.

In Nishizaki and Sakawa [61], there is small discrepancy in the sense that η is taken as λ rather than $(1-\lambda)$. Because of this, it is also mentioned there that the two linear programming problems corresponding to Player I's and Player II's problems are dual to each other in crisp sense. This is obviously not true as explained below in Section 6.4.

6.4 Special cases

It has already been explained in Remark 6.2.3 that various results of crisp two person zero sum matrix game theory follow as a special case of the fuzzy matrix game *FG*. In this section certain other special cases are presented so as to bring out the clear cut differences and similarities between the generalized model discussed in Section 6.2 and those presented here in Section 6.3.

Case 1

Let us use the notation of Section 6.3 so as to take $v_0=\bar{a}, w_0=\underline{a}, p_0=(\bar{a}-\underline{a})$, i.e. $\bar{a}$ (respectively $\underline{a}$) is the aspiration level of Player I (respectively Player II) and both the Players have the same tolerance level namely, $(\bar{a} - \underline{a})$. This later assumption, that both players have the same tolerance level, $(\bar{a} - \underline{a})$ appears to be rather restrictive. Now for this special case, (*FLP*) and (*FLD*) reduce to

(*FLP*1) max λ

subject to,

$$\left(\frac{A_j^T}{\bar{a} - \underline{a}}\right)x - \frac{\bar{a}}{\bar{a} - \underline{a}} \geq \lambda, \ (j = 1, 2, \ldots, n),$$
$$e^T x = 1,$$
$$\lambda \leq 1,$$
$$x, \lambda \geq 0,$$

and

(*FLD*1) max η

subject to,

$$\left(\frac{A_i}{\bar{a} - \underline{a}}\right)y - \frac{\underline{a}}{\bar{a} - \underline{a}} \leq 1 - \eta, \ (i = 1, 2, \ldots, m),$$
$$e^T y = 1,$$
$$\eta \leq 1,$$
$$y, \eta \geq 0.$$

These problems (*FLP*1) and (*FLD*1) are precisely the same problems as obtained in Section 6.3 for the very special case when both players have the same tolerance level of $(\bar{a} - \underline{a})$. It may be pointed out here that the approach taken in Section 6.2 and 6.3 are essentially similar. However it is mentioned in Section 6.3 that problems (*FLP*1) and (*FLD*1) are dual to each other in the crisp sense, i.e., (*FLD*1) is the conventional dual of (*FLP*1) and vice-versa, obviously is not correct. Therefore in contrast with what has been mentioned in Section 6.3, the degrees of attainment of two Players can not be equal. In fact even for this restricted case if λ^* is the degree of attainment of Player I then the degree of attainment for Player II will be $\eta^* = 1 - \lambda^*$. This follows by calling $(1-\eta)$ as ξ and then converting the (*FLD*1) into the minimization form. As has been established in Section 6.2 (*FLP*1) and (*FLD*1), being the special case (*FLP*) and (*FLD*), constitute a fuzzy primal-dual pair of linear programming problems.

Now if (x^*, λ^*) and (y^*, η^*) are optimal solutions of (*FLP*1) and (*FLD*1) respectively then for $\lambda^* = \eta^* = 1$, the fuzzy matrix game of

Section 6.3 should reduce to the crisp matrix game G. In that case the constraints of ($FLP1$) and ($FLD1$) become

$$\sum_{i=1}^{m} a_{ij}x_i^* \geq \bar{a}, \ (j = 1, 2, \ldots, n),$$
$$\sum_{i=1}^{m} x_i^* = 1,$$
$$x^* \geq 0,$$

and

$$\sum_{j=1}^{n} a_{ij}y_j^* \leq \underline{a}, \quad (i = 1, 2, \ldots, m),$$
$$\sum_{j=1}^{n} y_j^* = 1,$$
$$y^* \geq 0,$$

which do not guarantee the fact that x^* and y^* are optimal strategies for Player I and Player II respectively. From the above constraints we see that this can happen only when $\bar{a} = \underline{a}$ = value of the game. In this case the tolerance $\bar{a} - \underline{a} = 0$, which is not possible and therefore ($FLP1$) and ($FLD1$) are not defined. Furthermore, if $\bar{a}$ and $\underline{a}$ are taken as

$$\bar{a} = \max_i \max_j \ a_{ij}$$

and

$$\underline{a} = \min_i \min_j \ a_{ij},$$

then $\bar{a} = \underline{a}$ implies that A is a constant matrix.

Therefore the fuzzy game formulation of Sakawa and Nishizaki [68] does not seem to capture the crisp scenario unless some modifications are made in it. The construction of the fuzzy matrix game FG as presented here seems to be the appropriate generalization for this purpose.

Case 2

If Player I and Player II specify the same aspiration level $v_0 = w_0 = \hat{v}$(say), then, irrespective of the fact whether p_0 and q_0 are same or different, the problems

(*FLP*2) max λ
subject to,

$$\lambda \leq 1 - \frac{(\hat{v} - A_j^T x)}{p_0}, (j = 1, 2, \ldots, n),$$
$$e^T x = 1,$$
$$\lambda \leq 1,$$
$$x, \lambda \geq 0,$$

and

(*FLD*2) max η
subject to,

$$\eta \leq 1 - \frac{(A_i y - \hat{v})}{q_0}, (i = 1, 2, \ldots, m),$$
$$e^T y = 1,$$
$$\eta \leq 1,$$
$$y, \eta \geq 0,$$

are still dual to each other in "fuzzy" sense.

These special cases are now illustrated through the following numerical examples.

Example 6.4.1. Consider the two person zero-sum crisp matrix game G whose pay-off matrix A is

$$A = \begin{pmatrix} 1 & 3 \\ 4 & 0 \end{pmatrix}.$$

It can be verified that the solution of the game G is (x^*, y^*, v^*) where $x^* = (2/3, 1/3)^T$, $y^* = (1/2, 1/2)^T$ and $v^* = 2$.

Next let us consider the fuzzy versions of the game G.

(i) Let $v_0 = w_0 = 2$, $p_0 = 1$, $q_0 = 2$. For this choice (*FLP*) and (*FLD*) come out to be

max λ
subject to,

$$-\lambda + x_1 + 4x_2 \geq 1$$
$$-\lambda + 3x_1 \geq 1$$
$$\lambda \leq 1$$
$$x_1 + x_2 = 1$$
$$\lambda, x_1, x_2 \geq 0,$$

and

$$\begin{aligned} \max \quad & \eta \\ \text{subject to,} \quad & \\ 2\eta + y_1 + 3y_2 &\leq 4 \\ \eta + 2y_1 &\leq 2 \\ \eta &\leq 1 \\ y_1 + y_2 &= 1 \\ \eta, y_1, y_2 &\geq 0, \end{aligned}$$

respectively.

The optimal solutions of these problems are $\left(x^* = (2/3, 1/3)^T,\ \lambda^* = 1\right)$ and $\left(y^* = (1/2, 1/2)^T,\ \eta^* = 1\right)$ respectively. As $v_0 = w_0$ and $\lambda^* = \eta^* = 1$, the situation is not really fuzzy and therefore FG coincides with G and the solutions of both the games are also same.

(ii) Let $v_0 = 5/2$, $w_0 = 3$, $p_0 = 1$ and $q_0 = 1/2$. This choice corresponds to the situation where Player I aspires to win more than 5/2 but is satisfied (with varying degree) if he wins more than 3/2. Similarly, Player II aspires not to loose more than 3 but he will be satisfied (with varying degree) if he loses at most 7/2. For this case (*FLP*) and (*FLD*) are

$$\begin{aligned} \max \quad & \lambda \\ \text{subject to,} \quad & \\ -\lambda + x_1 + 4x_2 &\geq 7/2 \\ -\lambda + 3x_1 &\geq 7/2 \\ \lambda &\leq 1 \\ x_1 + x_2 &= 1 \\ \lambda, x_1, x_2 &\geq 0, \end{aligned}$$

and

$$\begin{aligned} \max \quad & \eta \\ \text{subject to,} \quad & \\ \eta + 2y_1 + 6y_2 &\leq 7 \\ \eta + 8y_1 &\leq 7 \\ \eta &\leq 1 \\ y_1 + y_2 &= 1 \\ \eta, y_1, y_2 &\geq 0, \end{aligned}$$

respectively.

The optimal solutions of these linear programming problems are $\big(x^* =$

$(2/3, 1/3)^T$, $\lambda^* = 1/2\big)$ and $\big(y^* = (3/4, 1/4)^T$, $\eta^* = 1\big)$ respectively and the two problems are dual to each other in the "fuzzy" sense as has been discussed earlier.

Example 6.4.2. This example depicts the situation considered in Section 6.3. Let

$$A = \begin{pmatrix} 10 & 8 \\ 6 & 2 \end{pmatrix}$$

Then $v_0 = \bar{a} = 10$, $w_0 = \underline{a} = 2$, and $p_0 = \bar{a} - \underline{a} = q_0 = 8$. In this case (*FLP*) and (*FLD*) become

$$\begin{aligned} &\max \quad \lambda \\ &\text{subject to,} \\ &-8\lambda + 10x_1 + 6x_2 \geq 2 \\ &-8\lambda + 8x_1 + 2x_2 \geq 2 \\ &\lambda \leq 1 \\ &x_1 + x_2 = 1 \\ &\lambda, x_1, x_2 \geq 0, \end{aligned}$$

and

$$\begin{aligned} &\max \quad \eta \\ &\text{subject to,} \\ &8\eta + 10y_1 + 8y_2 \leq 10 \\ &8\eta + 6y_1 + 2y_2 \leq 10 \\ &\eta \leq 1 \\ &y_1 + y_2 = 1 \\ &\eta, y_1, y_2 \geq 0, \end{aligned}$$

respectively.

The optimal solutions of these problems are $\big(x^* = (1, 0)^T$, $\lambda^* = 3/4\big)$ and $\big(y^* = (0, 1)^T$, $\eta^* = 1/4)$ respectively. Here, in variance with what is mentioned in Nishizaki and Sakawa [61], neither λ^* and η^* are same nor the above two problems constitute the usual primal-dual pair of linear programming problems. Thus this example illustrates our discussion of Remark 6.2.2.

6.5 Conclusions

In this chapter we have presented two models for studying two person zero sum matrix games with fuzzy goals. The main result established here is the equivalence of such a game with a primal-dual pair of fuzzy linear programming problems.

Since, in general, the strategy spaces of Player I and Player II could be polyhedral sets, we may also conceptualize constrained fuzzy matrix game on the lines of crisp constrained matrix games discussed in Chapter 1. Some relevant references in this direction are Li ([40], [41]) and Vijay [76].

7

Matrix games with fuzzy pay-offs

7.1 Introduction

In the last chapter, we have discussed matrix games with fuzzy goals and, based on duality in fuzzy linear programming, outlined a procedure to find a solution for the same. The main result proved there, gives complete equivalence of such a game with a suitable pair of primal-dual fuzzy linear programming problems.

The next topic to study in this sequel is *matrix game with fuzzy payoffs* which has earlier been studied by Campos [10] and later extended by Nishizaki and Sakawa [61] for the multiobjective situation. In the literature there are many models of matrix games with fuzzy payoffs but most of them seem to have taken inspiration from Campos [10] only, as it still remains the most basic work on this topic. Here in this chapter, as well as in the next chapter we present three such models which deal with the same problem but from totally different approaches. Specifically, the first model (Bector, Chandra and Vijay [7]) uses a suitable defuzzification function to establish certain duality for linear programming with fuzzy parameters and employs the same to solve matrix games with fuzzy pay-offs. The proposed algorithm is essentially the same as that of Campos [10] but certain modifications are needed to justify various steps involved therein. These results, in a sense, complement/supplement the basic ideas of Campos [10] and help in having a better understanding of the same. The second model, called Maeda's model [50], is based on fuzzy max order to define three kinds minimax equilibrium strategies whose properties help in the development of a solution procedure for solving such games. The third and the last model is due to Li [39] which presents a multiobjective linear programming approach to solve such games when the elements of the pay-off matrix are triangular fuzzy numbers.

While this chapter is devoted to the study of the first model (Bector, Chandra and Vijay [7]) the other two models, namely, Maeda's model [50] and Li's model [39] will be discussed in Chapter 8. We shall also attempt to extend some of these models so as to take into consideration the matrix games with fuzzy goals as well as fuzzy pay-offs.

This chapter has five main sections, namely; *definitions and preliminaries*, *duality in linear programming with fuzzy parameters*, *two person zero sum matrix game with fuzzy pay-offs: a defuzzification function approach*, *Campos' model: some comments*, and *matrix game with fuzzy goals and fuzzy pay-offs via a defuzzification function approach.*

7.2 Definitions and preliminaries

In addition to various definitions and preliminaries as introduced in Section 1.3 with regard to the crisp matrix game $G = (S^m,\ S^n,\ A)$ we need to understand the concept of double fuzzy constraints, i.e., constraints which are expressed as fuzzy inequalities involving fuzzy numbers. For this, let $N(\mathbb{R})$ be the set of all fuzzy numbers. Also let $\tilde{A}$, $\tilde{b}$ and $\tilde{c}$ respectively be $(m \times n)$ matrix, $(m \times 1)$ and $(n \times 1)$ vector having entries from $N(\mathbb{R})$, and the double fuzzy constraints under consideration be given by $\tilde{A}x \lesssim_{\tilde{p}} \tilde{b}$ and $\tilde{A}^T y \gtrsim_{\tilde{q}} \tilde{c}$, with adequacies $\tilde{p}$ and $\tilde{q}$ respectively.

Based on a resolution method proposed in [87], the constraint $\tilde{A}x \lesssim_{\tilde{p}} \tilde{b}$ is expressed as $\tilde{A}x \ \circledleq\ \tilde{b} + \tilde{p}(1 - \lambda),\quad \lambda \in [0, 1]$ where for $i = (1, 2, \ldots, m)$ the ith component of the fuzzy vector $\tilde{p}$, namely $\tilde{p}_i$, measures the adequacy between the fuzzy numbers $\tilde{A}_i x$ and $\tilde{b}_i$ which are the ith component of fuzzy vectors $\tilde{A}x$ and $\tilde{b}$ respectively. Similarly, the constraint $\tilde{A}^T y \gtrsim_{\tilde{q}} \tilde{c}$ is expressed as $\tilde{A}^T y \ \circledgeq\ \tilde{c} - \tilde{q}(1 - \eta),\quad \eta \in [0, 1]$, where for $j = (1, 2, \ldots, n)$ the jth component of the fuzzy vector $\tilde{q}$, namely $\tilde{q}_j$, measures the adequacy between the fuzzy numbers $\tilde{A}_j^T y$ and $\tilde{c}_j$ which are the jth component of fuzzy vectors $\tilde{A}^T y$ and $\tilde{c}$ respectively. Here $\circledleq$ and $\circledgeq$ are relations between fuzzy numbers which preserve the ranking when fuzzy numbers are multiplied by positive scalars. For example, this could be with respect to any ranking function $F : N(\mathbb{R}) \to \mathbb{R}$ taken in Campos [10] such that $\tilde{a} \circledleq \tilde{b}$ implies $F(\tilde{a}) \leq F(\tilde{b})$. There is also an implicit additional assumption of linearity of F in Campos [10] which is being taken here as well. Since in subsequent sections, the function F is used to defuzzify the given fuzzy linear programming problems, here onwards it is called as *defuzzification function* rather than a ranking function.

Therefore, the double fuzzy constraints of the type $\tilde{A}x \lesssim_{\tilde{p}} \tilde{b}$ and $\tilde{A}^T y \gtrsim_{\tilde{q}} \tilde{c}$ are to be understood as

$$\tilde{A}_i x \circledleq \tilde{b}_i + (1-\lambda)\tilde{p}_i \text{ for } 0 \leq \lambda \leq 1 \text{ and } (i = 1,2,\ldots,m),$$

and

$$\tilde{A}_j^T y \circledgeq \tilde{c}_j - (1-\eta)\tilde{q}_j \text{ for } 0 \leq \eta \leq 1 \text{ and } (j = 1,2,\ldots,n);$$

which in turn means

$$F(\tilde{A}_i x) \leq F(\tilde{b}_i) + (1-\lambda)F(\tilde{p})$$

and

$$F(\tilde{A}_j^T y) \geq F(\tilde{c}_j) - (1-\eta)F(\tilde{q}).$$

Now, let $\tilde{a}_{ij}$, $\tilde{b}_i$, $\tilde{p}_i$, $\tilde{c}_j$ and $\tilde{q}_j$ are triangular fuzzy numbers (TFNs) and F is the Yager's [87] first index given by

$$F(D) = \frac{\int_{d_l}^{d_u} x\mu_D(x)dx}{\int_{d_l}^{d_u} \mu_D(x)dx},$$

where d_l and d_u are the lower limits and upper limits of the support of the fuzzy number D. Then for the special case of TFNs the constraints $\tilde{A}x \lesssim_{\tilde{p}} \tilde{b}$ and $\tilde{A}^T y \gtrsim_{\tilde{q}} \tilde{c}$ respectively mean

$$\sum_{j=1}^{n} \Big((a_{ij})_l + a_{ij} + (a_{ij})_u\Big)x_j \leq \Big((b_i)_l + b_i + (b_i)_u\Big) + (1-\lambda)\Big((p_i)_l + p_i + (p_i)_u\Big)$$

and

$$\sum_{i=1}^{m} \Big((a_{ij})_l + a_{ij} + (a_{ij})_u\Big)y_i \geq \Big((c_j)_l + c_j + (c_j)_u\Big) - (1-\eta)\Big((q_j)_l + q_j + (q_j)_u\Big)$$

for $\lambda \in [0,1]$, $\eta \in [0,1]$, $i = 1,\ldots,m$ and $j = 1,\ldots,n$.

Here $\tilde{a}_{ij} = \Big((a_{ij})_l, a_{ij}, (a_{ij})_u\Big)$, $\tilde{b}_l = \Big((b_i)_l, b_i, (b_i)_u\Big)$, $\tilde{p}_i = \Big((p_i)_l, p_i, (p_i)_u\Big)$, $\tilde{c}_j = \Big((c_j)_l, c_j, (c_j)_u\Big)$ and $\tilde{q}_j = \Big((q_j)_l, q_j, (q_j)_u\Big)$ are TFNs.

7.3 Duality in linear programming with fuzzy parameters

Taking motivation from the usual crisp pair of primal-dual linear programming problems, we consider a very natural fuzzy version of the usual primal and dual problems as given below and explain their meaning. Specifically these problems are

(*FP*1)
$$\begin{aligned} \max \quad & \tilde{c}^T x \\ \text{subject to,} \quad & \\ & \tilde{A}x \lesssim \tilde{b}, \\ & x \geq 0, \end{aligned}$$

and

(*FD*1)
$$\begin{aligned} \min \quad & \tilde{b}^T y \\ \text{subject to,} \quad & \\ & \tilde{A}^T y \gtrsim \tilde{c}, \\ & y \geq 0. \end{aligned}$$

Here, $\tilde{A}$ is an $(m \times n)$ matrix of fuzzy numbers, and $\tilde{b}$ and $\tilde{c}$ respectively are $(m \times 1)$ and $(n \times 1)$ vectors of fuzzy numbers. The symbols '$\lesssim$' and '$\gtrsim$' are fuzzy versions of the symbols '$\leq$' and '$\geq$' respectively, and have the linguistic interpretation "essentially less than or equal to" and "essentially greater than or equal to" as explained in Zimmermann [90]. Also, the double fuzzy constraint $\tilde{A}x \lesssim \tilde{b}$ and $\tilde{A}^T y \gtrsim \tilde{c}$ are to be understood with respect to a suitable defuzzification function F and adequacies $\tilde{p}$ and $\tilde{q}$, in the sense as explained in Section 7.2. It should further be noted that the defuzzification function F once chosen is to be kept fixed for all development in this sequel. Therefore, if $F : N(\mathbb{R}) \to \mathbb{R}$ is the chosen defuzzification function of fuzzy numbers for constraints in (*FP*1) and (*FD*1) then utilizing the same defuzzification function F for the objective functions in (*FP*1) and (*FD*1), we get (*FP*2) and (*FD*2) as follows

(*FP*2)
$$\begin{aligned} \max \quad & F(\tilde{c}^T x) \\ \text{subject to,} \quad & \\ & F(\tilde{A}x) \leq F(\tilde{b}) + (1 - \lambda)F(\tilde{p}), \\ & \lambda \leq 1, \\ & x, \lambda \geq 0, \end{aligned}$$

and

(*FD*2)
$$\begin{aligned} \min \quad & F(\tilde{b}^T y) \\ \text{subject to,} \quad & \\ & F(\tilde{A}^T y) \geq F(\tilde{c}) - (1 - \eta)F(\tilde{q}), \\ & \eta \leq 1, \\ & y, \eta \geq 0. \end{aligned}$$

Here $\tilde{p}$ and $\tilde{q}$ respectively measure the adequacies in the primal and dual constraints as explained earlier.

The pair (*FP*2) and (*FD*2) is termed as *the fuzzy pair of primal-dual linear programming problems.*

We shall now prove the following modified weak duality theorem for the pair (*FP*2) and (*FD*2).

Theorem 7.3.1 *Let* $(x,\ \lambda)$ *be* (*FP*2)*-feasible and* $(y,\ \eta)$ *be* (*FD*2)*-feasible. Then*

$$F(\tilde{c}^T x) - F(\tilde{b}^T y) \leq (1-\lambda)F(\tilde{p}^T y) + (1-\eta)F(\tilde{q}^T x).$$

Proof. Since $(x,\ \lambda)$ is (*FP*2)-feasible and $(y,\ \eta)$ is (*FD*2)-feasible, we have

$$F(\tilde{A}x) \leq F(\tilde{b}) + (1-\lambda)F(\tilde{p}),\ x \geq 0,$$

and

$$F(\tilde{A}^T y) \geq F(\tilde{c}) - (1-\eta)F(\tilde{q}),\ y \geq 0.$$

Now because of the properties of relations $\lesssim$ and $\gtrsim$ the defuzzification function F preserves the ranking when fuzzy numbers are multiplied by non-negative scalars, the above relations imply

$$F(x^T \tilde{A}^T y) \leq F(\tilde{b}^T y) + (1-\lambda)F(\tilde{p}^T y),$$

and

$$F(y^T \tilde{A}x) \geq F(\tilde{c}^T x) - (1-\eta)F(\tilde{q}^T x).$$

Therefore,

$$F(\tilde{b}^T y) + (1-\lambda)F(\tilde{p}^T y) \geq F(\tilde{c}^T x) - (1-\eta)F(\tilde{q}^T x),$$

because

$$F(x^T \tilde{A}^T y) = F(y^T \tilde{A}x),$$

as

$$x^T \tilde{A}^T y = y^T \tilde{A}x.$$

Combining the above, we obtain

$$F(\tilde{b}^T y) - F(\tilde{c}^T x) \geq (\lambda - 1)F(\tilde{p}^T y) + (\eta - 1)F(\tilde{q}^T x).$$

Remark 7.3.1. In case $\tilde{A}$, $\tilde{c}$ and $\tilde{b}$ are crisp and $\lambda = 1$ and $\eta = 1$ then the pair (*FP*2)-(*FD*2) reduces to the usual crisp primal-dual pair and Theorem 7.3.1 becomes the usual weak duality theorem.

Remark 7.3.2. In a very recent work, Inuiguchi et al. [29] studied fuzzy linear programming duality in the setting of "fuzzy (valued) relations". The approach described above is different from that of [29] as here a defuzzification function F is used and results are stated in terms of this function only. In Chapter 10, we shall have opportunity to discuss the approach of Inuiguchi et al. [29] along with some other recent approaches for studying fuzzy duality e.g. Wu [82].

The above discussion on duality can further be extended to include even those linear programming problems in which both the parameters as well as the constraints are fuzzy. For this, we consider the following pair of fuzzy linear programming problems

(*GFP*1) Find $x \in R^n$ such that

$$\begin{aligned} \tilde{c}^T x &\gtrsim \tilde{Z}_0, \\ \tilde{A}x &\lesssim \tilde{b}, \\ x &\geq 0, \end{aligned}$$

and,

(*GFD*1) Find $y \in R^m$ such that

$$\begin{aligned} \tilde{b}^T y &\lesssim \tilde{W}_0, \\ \tilde{A}^T y &\gtrsim \tilde{c}, \\ y &\geq 0. \end{aligned}$$

Here $\tilde{Z}_0$ and $\tilde{W}_0$ are fuzzy quantities which represent aspiration levels for the objective functions in problems (*GFP*1) and (*GFD*1) respectively. Further, $\tilde{p}_0$ (respectively $\tilde{q}_0$) measures the adequacy between the objective function $\tilde{c}^T x$ (respectively $\tilde{b}^T y$) and the aspiration level $\tilde{Z}_0$ (respectively $\tilde{W}_0$). Rest of the symbols and notations have the same meaning as explained earlier in the context of problems (*FP*1) and (*FD*1) in Section 7.3.

Now, if $F : N(\mathbb{R}) \to \mathbb{R}$ is the chosen defuzzification function of fuzzy numbers for constraints in (*GFP*1) and (*GFD*1), we get (*GFP*2) and (*GFD*2) as

(*GFP*2) max λ

subject to,

$$\begin{aligned} F(\tilde{c}^T x) &\geq F(\tilde{Z}_0) - F(\tilde{p}_0)(1-\lambda), \\ F(\tilde{A}_i x) &\leq F(\tilde{b}_i) + F(\tilde{p}_i)(1-\lambda), \quad (i = 1, 2, \ldots, m), \\ \lambda &\leq 1, \\ x, \lambda &\geq 0, \end{aligned}$$

and

(*GFD*2) max η

subject to,

$$F(\tilde{b}^T y) \leq F(\tilde{W}_0) + F(\tilde{q}_0)(1-\eta)$$
$$F(\tilde{A}_j^T y) \geq F(\tilde{c}_j) - F(\tilde{q}_j)(1-\eta), \quad (j = 1, 2, \ldots, n),$$
$$\eta \leq 1,$$
$$y, \eta \geq 0,$$

which could further be written as follows

(*GFP*3) max λ

subject to,

$$\lambda \leq 1 + \frac{F(\tilde{c}^T)x - F(\tilde{Z}_0)}{F(\tilde{p}_0)},$$
$$\lambda \leq 1 + \frac{F(\tilde{b}_i) - F(\tilde{A}_i)x}{F(\tilde{p}_i)}, \quad (i = 1, 2, \ldots, m),$$
$$\lambda \leq 1,$$
$$x, \lambda \geq 0,$$

and

(*GFD*3) min $(-\eta)$

subject to,

$$\eta \leq 1 + \frac{F(\tilde{W}_0) - F(\tilde{b}^T)y}{F(\tilde{q}_0)},$$
$$\eta \leq 1 + \frac{F(\tilde{A}_j^T)y - F(\tilde{c}_j)}{F(\tilde{q}_j)}, \quad (j = 1, 2, \ldots, n),$$
$$\eta \leq 1,$$
$$y, \eta \geq 0.$$

The pair (*GFP*3) and (*GFD*3) is termed as the fuzzy pair of primal-dual linear programming problems in the sense of Bector and Chandra [4] as discussed in Chapter 5. Let $F(\tilde{p}) = \left(F(\tilde{p}_1), F(\tilde{p}_2), \ldots, F(\tilde{p}_m)\right)^T$ and $F(\tilde{q}) = \left(F(\tilde{q}_1), F(\tilde{q}_2), \ldots, F(\tilde{q}_n)\right)^T$. Then the following duality theorem follows from Bector and Chandra [4].

Theorem 7.3.2 *Let* (x, λ) *be* (*GFP*3)*-feasible and* (y, η) *be* (*GFD*3)*-feasible. Then* $(\lambda - 1)F(\tilde{p})^T y + (\eta - 1)F(\tilde{q})^T x \leq \left(F(\tilde{b})^T y - F(\tilde{c})^T x\right)$.

7.4 Two person zero sum matrix games with fuzzy pay-offs: main results

Let S^m, S^n be as introduced in Section 1.3 and $\tilde{A}$ be the pay-off matrix with entries as fuzzy numbers. Then *a two person zero-sum matrix game with fuzzy pay-offs* is the triplet

$$FG = (S^m,\ S^n,\ \tilde{A}).$$

In the following, we shall often call a two person zero-sum matrix game with fuzzy pay-offs simply as fuzzy matrix game. Now, we define the meaning of the *solution* of the fuzzy matrix game *FG*.

Definition 7.4.1 (Reasonable solution of the game *FG***).** *Let* $\tilde{v}, \tilde{w} \in N(R)$*. Then* $(\tilde{v},\ \tilde{w})$ *is called a reasonable solution of the fuzzy matrix game FG if there exists* $x^* \in S^m$*,* $y^* \in S^n$ *satisfying*

$$(x^*)^T \tilde{A} y \gtrsim \tilde{v},\ \forall y \in S^n$$

and

$$x^T \tilde{A} y^* \lesssim \tilde{w},\ \forall x \in S^m.$$

If $(\tilde{v},\ \tilde{w})$ *is a reasonable solution of FG then* $\tilde{v}$ *(respectively* $\tilde{w}$*) is called a reasonable value for Player I (respectively Player II).*

Definition 7.4.2 (Solution of the game *FG***).** *Let* T_1 *and* T_2 *be the set of all reasonable values* $\tilde{v}$ *and* $\tilde{w}$ *for Player I and Player II respectively where* $\tilde{v}, \tilde{w} \in N(R)$*. Let there exist* $\tilde{v}^* \in T_1$*,* $\tilde{w}^* \in T_2$ *such that*

$$F(\tilde{v}^*) \geq F(\tilde{v})\ ,\ \forall \tilde{v} \in T_1$$

and

$$F(\tilde{w}^*) \leq F(\tilde{w})\ ,\ \forall \tilde{w} \in T_2.$$

Then $(x^*,\ y^*,\ \tilde{v}^*,\ \tilde{w}^*)$ *is called the solution of the game FG where* $\tilde{v}^*$ *(respectively* $\tilde{w}^*$*) is the value of the game FG for Player I (respectively Player II) and* x^* *(respectively* y^**) is called an optimal strategy for Player I (respectively Player II).*

By using the above definitions for the game *FG*, we now construct the following pair of fuzzy linear programming problems for Player I and Player II

(*FP*3)

$$\begin{aligned} \max \quad & F(\tilde{v}) \\ \text{subject to,} \quad & \\ & x^T \tilde{A} y \gtrsim_{\tilde{p}} \tilde{v}, \ \text{ for all } y \in S^n, \\ & x \in S^m, \end{aligned}$$

and

(*FD*3)

$$\min \quad F(\tilde{w})$$

subject to,

$$x^T \tilde{A} y \lesssim_{\tilde{q}} \tilde{w}, \text{ for all } x \in S^m,$$
$$y \in S^n.$$

Now recalling the explanation of the double fuzzy constraints as explained in Section 7.2 and noting that the relations $\textcircled{\scriptsize$\geq$}$ and $\textcircled{\scriptsize$\leq$}$ preserve the ranking when fuzzy numbers are multiplied by positive scalars, it makes sense to consider only the extreme points of sets S^m and S^n in the constraints of $(FP3)$ and $(FD3)$. Therefore the above problems $(FP3)$ and $(FD3)$ will be converted into

$(FP4)$ max $F(\tilde{v})$

subject to,

$$x^T \tilde{A}_j \gtrsim_{\tilde{p}} \tilde{v}, (j = 1, 2, \ldots, n),$$
$$e^T x = 1,$$
$$x \geq 0,$$

and

$(FD4)$ min $F(\tilde{w})$

subject to,

$$\tilde{A}_i y \lesssim_{\tilde{q}} \tilde{w}, (i = 1, 2, \ldots, m),$$
$$e^T y = 1,$$
$$y \geq 0.$$

Here $\tilde{A}_i$ (respectively $\tilde{A}_j$) denotes the ith row (respectively j^{th} column) of $\tilde{A}$ $(i = 1, 2, \cdots, m;\ j = 1, 2, \cdots, n)$.

By using the resolution procedure for the double fuzzy constraints in $(FP4)$ and $(FD4)$, we obtain

$(FP5)$ max $F(\tilde{v})$

subject to,

$$\sum_{i=1}^{m} \tilde{a}_{ij} x_i \textcircled{\scriptsize$\geq$} \tilde{v} - (1 - \lambda)\tilde{p}, (j = 1, 2, \ldots, n),$$
$$e^T x = 1,$$
$$\lambda \leq 1,$$
$$x, \lambda \geq 0,$$

and

$(FD5)$ min $F(\tilde{w})$

subject to,

$$\sum_{j=1}^{n} \tilde{a}_{ij} y_j \lesssim \tilde{w} + (1-\eta)\tilde{q}, \ (i = 1, 2, \ldots, m),$$
$$e^T y = 1,$$
$$\eta \leq 1,$$
$$y, \eta \geq 0,$$

Now by utilizing the defuzzification function $F : N(R) \to R$ for the constraints ($FP5$) and ($FD5$), these problems can further be written as

($FP6$) max $F(\tilde{v})$

subject to,

$$\sum_{i=1}^{m} F(\tilde{a}_{ij}) x_i \geq F(\tilde{v}) - (1-\lambda) F(\tilde{p}), \ (j = 1, 2, \ldots, n),$$
$$e^T x = 1,$$
$$\lambda \leq 1,$$
$$x, \lambda \geq 0,$$

and

($FD6$) min $F(\tilde{w})$

subject to,

$$\sum_{j=1}^{n} F(\tilde{a}_{ij}) y_j \leq F(\tilde{w}) + (1-\eta) F(\tilde{q}), \ (i = 1, 2, \ldots, m),$$
$$e^T y = 1,$$
$$\eta \leq 1,$$
$$y, \eta \geq 0.$$

From the above discussion we observe that for solving the fuzzy matrix game FG we have to solve the crisp linear programming problems ($FP6$) and ($FD6$) for Player I and Player II respectively. Also, if $(x^*, \lambda^*, \tilde{v}_*)$ is an optimal solution of ($FP6$) then for Player I, x^* is an optimal strategy, $\tilde{v}_*$ is the fuzzy value and $(1-\lambda^*)\tilde{p}$ is the measure of the adequacy level for the double fuzzy constraints in ($FP5$). Similar interpretation can also be given to an optimal solution $(y^*, \eta^*, \tilde{w}_*)$ of the problem ($FD6$). Further the results of Section 7.3 show that for the pair ($FP6$) and ($FD6$) the following theorem holds:

Theorem 7.4.1 *The pair* ($FP6$)-($FD6$) *constitutes a fuzzy primal-dual pair in the sense of Theorem 7.3.1.*

All the results discussed in this section can now be summarized in the form of Theorem 7.4.2 given below.

Theorem 7.4.2 *The fuzzy matrix game FG described by FG* = (S^m, S^n, $\tilde{A}$) *is equivalent to two crisp linear programming problems* (*FP*6) *and* (*FD*6) *which constitute a primal-dual pair in the sense of duality for linear programming with fuzzy parameters.*

Remark 7.4.1. It is important to note that the crisp problems (*FP*6) and (*FD*6) do not constitute a primal-dual pair in the conventional sense of duality in linear programming but are dual in "fuzzy" sense as explained above. Therefore if $(x^*, \lambda^*, \tilde{v}_*)$ is optimal to (*FP*6) and $(y^*, \eta^*, \tilde{w}_*)$ is optimal to (*FD*6) then in general one should not expect that $F(\tilde{v}_*) = F(\tilde{w}_*)$.

Remark 7.4.2. If all the fuzzy numbers are to be taken as crisp numbers i.e. $\tilde{a}_{ij} = a_{ij}$, $\tilde{b}_i = b_i$, $\tilde{c}_j = c_j$ and in the optimal solutions of (*FP*6) and (*FD*6), $\lambda^* = \eta^* = 1$, then the fuzzy game *FG* reduces to the crisp two person zero sum game *G*. Thus if $\tilde{A}$, $\tilde{b}$, $\tilde{c}$ are crisp numbers and $\lambda^* = \eta^* = 1$, *FG* reduces to *G*; the pair (*FP*6)-(*FD*6) reduces to the pair (*LP*)-(*LD*); and as it should be, Theorem 7.4.2 reduces to Theorem 1.4.1.

Remark 7.4.3. In general it may be difficult to obtain exact membership functions for fuzzy values $\tilde{v}_*$ and $\tilde{w}_*$ because of the large number of parameters involved in their representation. For example, if $\tilde{v}$ is a TFN (v_l, v, v_u) then to determine $\tilde{v}$ completely we need all of these three variables. Therefore, purely from the computational point of view it becomes easier to take $F(\tilde{v})$ and $F(\tilde{w})$ as real variables V and W respectively and modify problems (*FP*6) and (*FD*6) as follows

(*FP*7) max V

subject to,

$$\sum_{i=1}^{m} F(\tilde{a}_{ij})x_i \geq V - (1-\lambda)F(\tilde{p}),\ (j = 1, 2, \ldots, n),$$
$$e^T x = 1,$$
$$\lambda \leq 1,$$
$$x, \lambda \geq 0,$$

and

(*FD*7) min W

subject to,

$$\sum_{j=1}^{n} F(\tilde{a}_{ij})y_j \leq W + (1-\eta)F(\tilde{q}), \ (i = 1, 2, \ldots, m),$$
$$e^T y = 1,$$
$$\eta \leq 1,$$
$$y, \eta \geq 0.$$

In this situation, in-spite of knowing that the value for Player I (respectively Player II) is fuzzy with certain membership function, we shall only get numerical values V^* (respectively W^*) for Player I (respectively Player II) and the actual fuzzy value for Player I and Player II will be "close to" V^* and W^* respectively. Thus we shall not get exact membership functions for the fuzzy values of Player I and Player II even though these are very much desirable. In the particular case when F is Yager's first index [87], the numerical values V^* (respectively W^*) will represent the "centroid" or "average" value for Player I (respectively Player II).

7.5 Campos' model: some comments

Campos [10] also considered the fuzzy game model $FG = (S^m, S^n, \tilde{A})$ earlier and taking motivation from the crisp case, suggested following linear programming problems ($FP8$) and ($FD8$) for Player I and Player II respectively

($FP8$)

$$\max \quad v$$
subject to,
$$\sum_{i=1}^{m} \tilde{a}_{ij}x_i \gtrsim v, \ (j = 1, 2, \ldots, n),$$
$$e^T x = 1,$$
$$x \geq 0,$$

and

($FD8$)

$$\min \quad w$$
subject to,
$$\sum_{j=1}^{n} \tilde{a}_{ij}y_j \lesssim w, \ (i = 1, 2, \ldots, m),$$
$$e^T y = 1,$$
$$y \geq 0,$$

where $v, w \in R$ and the double fuzzy inequalities in $(FP8)$ and $(FD8)$ are to be understood as discussed here in Section 7.2.

Further, following a parallel way to the classical crisp case, Campos [10] argued that v, w can be taken to be strictly positive. Therefore, one can define $u \in R^m_+$, $s \in R^n_+$ such that $u_i = \dfrac{x_i}{v}$ $(i = 1, 2, \ldots, m)$ and $s_j = \dfrac{y_j}{w}$ $(j = 1, 2, \ldots, n)$ and that gives

$$v = \frac{1}{\sum_{i=1}^{m} u_i}, \; w = \frac{1}{\sum_{j=1}^{n} s_j}.$$

Also, then problems $(FP8)$ and $(FD8)$ respectively get transferred to,

$$(FP9) \qquad \min \quad \sum_{i=1}^{m} u_i$$

subject to,

$$\sum_{i=1}^{m} \tilde{a}_{ij} u_i \gtrsim 1, \; (j = 1, \ldots, n),$$
$$u_i \geq 0, \; (i = 1, \ldots, m),$$

and

$$(FD9) \qquad \max \quad \sum_{j=1}^{n} s_j$$

subject to,

$$\sum_{j=1}^{n} \tilde{a}_{ij} s_j \lesssim 1, \; (i = 1, \ldots, m),$$
$$s_j \geq 0, \; (j = 1, \ldots, n).$$

Now expressing the double fuzzy constraints in $(FP9)$ and $(FD9)$ in terms of adequacies $\tilde{p}$ and $\tilde{q}$ as described in Section 7.2, we can rewrite these problems as

$$(FP10) \qquad \min \quad \sum_{i=1}^{m} u_i$$

subject to,

$$\sum_{i=1}^{m} \tilde{a}_{ij} u_i \; \textcircled{\scriptsize$\gtrsim$} \; 1 - (1 - \alpha) \tilde{p}_j, \; (j = 1, \ldots, n),$$
$$u_i \geq 0, \qquad (i = 1, \ldots, m),$$
$$\alpha \in [0, 1],$$

and

$$(FD10) \qquad \max \quad \sum_{j=1}^{n} s_j$$

subject to,

$$\sum_{j=1}^{n} \tilde{a}_{ij} s_j \lessapprox 1 + (1-\beta)\tilde{q}_i, \; (i = 1, \ldots, m),$$
$$s_j \geq 0, \qquad (j = 1, \ldots, n),$$
$$\beta \in [0,1],$$

where the scalar 1 in the right hand side of $(FP10)$ and $(FD10)$ is to be taken as the fuzzy number 1 i.e. $b_i = (b_i)_l = (b_i)_u = 1$ for all $i = 1, 2, \ldots, m$.

Now, in case $\tilde{a}_{ij}$, $\tilde{p}_i$, $\tilde{q}_j$ are triangular fuzzy numbers and Yager's [87] index is used for the relation $\lessapprox$ and $\gtrapprox$ in the above problems then $(FP10)$ and $(FD10)$ reduce to

$$(FP11) \quad \min \quad \sum_{i=1}^{m} u_i$$

subject to,

$$\sum_{i=1}^{m} \Big((\tilde{a}_{ij})_l + \tilde{a}_{ij} + (\tilde{a}_{ij})_u \Big) u_i \geq 3 - \Big((p_j)_l + p_j + (p_j)_u \Big)(1-\alpha), \; (j = 1, \ldots, n),$$
$$u_i \geq 0, \qquad (i = 1, \ldots, m),$$
$$\alpha \in [0,1],$$

and

$$(FD11) \quad \max \quad \sum_{j=1}^{n} s_j$$

subject to,

$$\sum_{j=1}^{n} \Big((\tilde{a}_{ij})_l + \tilde{a}_{ij} + (\tilde{a}_{ij})_u \Big) s_j \leq 3 + \Big((q_i)_l + q_i + (q_i)_u \Big)(1-\beta), \; (i = 1, \ldots, m),$$
$$s_j \geq 0, \qquad (j = 1, \ldots, n),$$
$$\beta \in [0,1].$$

Looking at the above development and other results mentioned in Campos [10] we make the following observations for the Campos' model

(i) In this model, though the pay-off matrix is fuzzy, as its elements are fuzzy numbers, the value of the game for Player I and Player II, namely v and w, are assumed to be crisp numbers. This is certainly obvious from the fact that in (*FP*8) and (*FD*8) v and w respectively are being minimized and maximized and later to get (*FP*9) and (*FD*9) there is division by v and w to get u_i, $(i = 1,2,\ldots,m)$ and s_j, $(j = 1,2,\ldots,n)$. Purely from the logical point of view it seems natural that if the pay-off matrix is fuzzy then the values for Player I and Player II should also be fuzzy, an argument that has been followed by Werners [79] in the context of fuzzy linear programming and discussed earlier in Chapter 4 In fact later Campos [10] also mentions that the value of the game $FG = (S^m, S^n, A)$ will be fuzzy and it will be around $v(1)$ and $w(1)$.
Thus in Campos [10] model if one takes v and w as fuzzy numbers then problems (*FP*8) and (*FD*8) can not be given any meaning as such and also the division by v and w to get u_i and s_j $(i = 1,2,\ldots,m;\ j = 1,2,\ldots,n)$ becomes meaningless. Our effort here is to start with fuzzy values for Player I and Player II and make appropriate modifications in (*FP*8) and (*FD*8) so that these problems become meaningful in both physical and mathematical terms. However, as noted in Remark 7.4.3, in actual practice one may not be able to get exact membership function for fuzzy values and be satisfied with representative values $F(\tilde{v})$ and $F(\tilde{w})$.

(ii) There is no justification to assume that $v,\ w > 0$ except that it is similar to the crisp situation. In crisp situation it is true because if v is the value of the game for the matrix A then $v+\alpha$ is the value of the game for the matrix $A(\alpha) = (a_{ij} + \alpha)$. It is not very clear if this happens for the fuzzy case as well. In fact as such, there seems to be no easy way to check it because a formal definition of the value of the fuzzy game is not given in Campos [10].

(iii) The constraints $\sum_{i=1}^{m} \tilde{a}_{ij} x_i \gtrsim v$ are simply written as $\sum_{i=1}^{m} \tilde{a}_{ij} u_i \gtrsim 1$ by dividing both sides with $v > 0$. This is correct if the constraints are crisp but may not be true if the constraints are fuzzy. It will be very much dependent upon the membership function and tolerance chosen. For example, if $x \gtrsim_p a$ denotes the fuzzy relation that x is "essentially more than a" with tolerance p and $\alpha > 0$, then for the linear membership function, it gives $\alpha x \gtrsim_{\alpha p} \alpha a$ and not $\alpha x \gtrsim_p \alpha a$.

(iv) The fuzzy linear programming problems (*FP*10) and (*FD*10) as obtained in Campos [10], do not constitute a pair of primal-dual problems in contrast with the usual crisp case. But as shown here in Section 7.4 if one starts with the correct conceptualization of fuzzy game then one can formulate a pair of linear programming problems which are dual to each other in fuzzy sense.

(v) In Campos' formulation, if we identify v and w as $F(\tilde{v})$ and $F(\tilde{w})$ respectively then the basic linear programming problems obtained there come out to be similar to (*FP*7) and (*FD*7) obtained here for the variables $V = F(\tilde{v})$ and $W = F(\tilde{w})$. But then the subsequent development does not seem to be correct in view of the observation (3) above.

Example 7.5.1. Consider the fuzzy game defined by the matrix of fuzzy numbers :

$$\mathbf{A} = \begin{bmatrix} \tilde{180} & \tilde{156} \\ \tilde{90} & \tilde{180} \end{bmatrix}$$

where $\tilde{180} = (175, 180, 190)$, $\tilde{156} = (150, 156, 158)$, $\tilde{90} = (80, 90, 100)$. Assuming that Player I and Player II have the margins $\tilde{p}_1 = \tilde{p}_2 = (0.08, 0.10, 0.11)$,
and $\tilde{q}_1 = \tilde{q}_2 = (0.14, 0.15, 0.17)$.

According to Theorem 7.4.2, to solve this game we have to solve following two crisp linear programming problems (*LP*1) and (*LD*1) for Player I and Player II respectively:

$$(LP1) \qquad \max \quad \frac{v_l + v + v_u}{3}$$

subject to,

$$\begin{aligned} 545x_1 + 270x_2 &\geq (v_l + v + v_u) - (1 - \lambda)(0.29) \\ 464x_1 + 545x_2 &\geq (v_l + v + v_u) - (1 - \lambda)(0.29) \\ x_1 + x_2 &= 1 \\ \lambda &\leq 1 \\ x_1, x_2, \lambda &\geq 0, \end{aligned}$$

and

$$(LD1) \qquad \min \quad \frac{w_l + w + w_u}{3}$$

subject to,

$$\begin{aligned} 545y_1 + 464y_2 &\leq (w_l + w + w_u) + (1 - \eta)(0.46) \\ 270y_1 + 545y_2 &\leq (w_l + w + w_u) + (1 - \eta)(0.46) \\ y_1 + y_2 &= 1 \\ \eta &\leq 1 \end{aligned}$$

$$y_1, y_2, \eta \geq 0.$$

Now to get the full membership representation of the fuzzy value for Player I (respectively Player II) one needs that in the optimal solution of $(LP1)$ (respectively $(LD1)$) all variables v_l, v, v_u (respectively w_l, w, w_u) come out to be non zero; i.e. they are basic variables. This seems to be most unlikely as there are much less number of constraints and therefore many of the variables are going to be non-basic and hence take zero values only. This observation motivates us to take $V = \dfrac{v_l + v + v_u}{3}$, $W = \dfrac{w_l + w + w_u}{3}$ and consider following problems $(LP2)$ and $(LD2)$ for the variables V and W

$(LP2)$ max V

subject to,

$$\begin{aligned} 545x_1 + 270x_2 &\geq 3V - (1-\lambda)(0.29) \\ 464x_1 + 545x_2 &\geq 3V - (1-\lambda)(0.29) \\ x_1 + x_2 &= 1 \\ \lambda &\leq 1 \\ x_1, x_2, \lambda &\geq 0, \end{aligned}$$

and

$(LD2)$ min W

subject to,

$$\begin{aligned} 545y_1 + 464y_2 &\leq 3W + (1-\eta)(0.46) \\ 270y_1 + 545y_2 &\leq 3W + (1-\eta)(0.46) \\ y_1 + y_2 &= 1 \\ \eta &\leq 1. \\ y_1, y_2, \eta &\geq 0. \end{aligned}$$

Solving the above Linear Programming Problems, we obtain, $(x_1^* = 0.7725,\ x_2^* = 0.2275,\ V = 160.91,\ \lambda^* = 0)$ and $(y_1^* = 0.2275,\ y_2^* = 0.7725,\ W = 160.65,\ \eta^* = 0)$.

Therefore we obtain optimal strategies for Player I and Player II as $(x_1^* = 0.7725,\ x_2^* = 0.2275)$ and $(y_1^* = 0.2275,\ y_2^* = 0.7725)$ respectively. Also, the fuzzy value of the game for Player I is "close to" 160.91. In a similar manner, the fuzzy value of the game for Player II is "close to" 160.65.

Here it may be noted that this solution of the given fuzzy game matches with that of Campos [10] though apparently different problems are being solved in [10]. This is basically because in this case one can assume that $V = F(\tilde{v}) = \dfrac{v_l + v + v_u}{3}$ and $W = F(\tilde{w}) = \dfrac{w_l + w + w_u}{3}$ are positive, and therefore by defining $u_1 = \dfrac{x_1}{F(\tilde{v})}, u_2 = \dfrac{x_2}{F(\tilde{v})}, s_1 = \dfrac{y_1}{F(\tilde{w})}$ and $s_2 = \dfrac{y_2}{F(\tilde{w})}$

problems ($LP2$) and ($LD2$) can be rewritten as

($LP3$)

$$\begin{aligned} \min \quad & u_1 + u_2 \\ \text{subject to,} \quad & \\ & 545u_1 + 270u_2 \geq 1 - \frac{(1-\lambda)(0.29)}{F(\tilde{v})} \\ & 464u_1 + 545u_2 \geq 1 - \frac{(1-\lambda)(0.29)}{F(\tilde{v})} \\ & \lambda \leq 1 \\ & u_1, u_2, \lambda \geq 0, \end{aligned}$$

and

($LD3$)

$$\begin{aligned} \max \quad & s_1 + s_2 \\ \text{subject to,} \quad & \\ & 545s_1 + 464s_2 \leq 1 - \frac{(1-\eta)(0.46)}{F(\tilde{w})} \\ & 270s_1 + 545s_2 \leq 1 - \frac{(1-\eta)(0.46)}{F(\tilde{w})} \\ & \eta \leq 1 \\ & s_1, s_2, \eta \geq 0. \end{aligned}$$

Now, following the arguments similar to Campos [10], it can be shown that solution of ($LP3$) and ($LD3$) will be obtained for $\lambda^* = 1$ and $\eta^* = 1$, the final result of ($LP3$) and ($LD3$) is bound to be the same as that of ($LP1$) and ($LD1$). Thus we may conclude that by the modifications as suggested here, the division can be performed to get problems of the type discussed in Campos [10]. It seems that in Campos [10], the division operation for the constraints is not done at the right place in the right manner and that creates some difficulty in getting the correct conceptualization of the corresponding linear programming problems for the given fuzzy game.

7.6 Matrix games with fuzzy goals and fuzzy payoffs

Let S^m, S^n be as introduced in Section 1.3 and $\tilde{A}$ be the pay-off matrix with entries as fuzzy numbers. Let $\tilde{v}$, $\tilde{w}$ be fuzzy numbers respectively the aspiration levels of Player I and Player II. Then a *two person zero-sum matrix game with fuzzy goals and fuzzy payoffs*, denoted by FG, is defined as:

$$FG = (S^m,\ S^n,\ \tilde{A},\ \tilde{v},\ \gtrsim,\ \tilde{p},\ \tilde{w},\ \lesssim, \tilde{q}),$$

where, '$\gtrsim$' and '$\lesssim$' have their meanings as explained in Section 7.2 and, $\tilde{p}$ and $\tilde{q}$ are fuzzy tolerance levels for Player I and Player II respectively.

Now onwards we shall call a two person zero-sum matrix game with fuzzy goals and fuzzy pay-offs simply as fuzzy matrix game.

We now define the meaning of the *solution* of the fuzzy matrix game *FG*.

Definition 7.6.1 *A point $(\bar{x}, \bar{y}) \in S^m \times S^n$ is called a solution of the fuzzy matrix game FG if*

$$(\bar{x})^T \tilde{A} y \gtrsim_{\tilde{p}} \tilde{v}, \text{ for all } y \in S^n$$

and,

$$x^T \tilde{A} \bar{y} \lesssim_{\tilde{q}} \tilde{w}, \text{ for all } x \in S^m.$$

Here $\bar{x}$ is called an optimal strategy for Player I and $\bar{y}$ is called an optimal strategy for Player II.

By using the above definitions for the game *FG*, we construct the following pair of fuzzy linear programming problems for Player I and Player II:

(*GFP*4) Find $x \in S^m$ such that,

$$x^T \tilde{A} y \gtrsim_{\tilde{p}} \tilde{v}, \text{ for all } y \in S^n,$$

and

(*GFD*4) Find $y \in S^n$ such that,

$$x^T \tilde{A} y \lesssim_{\tilde{q}} \tilde{w}, \text{ for all } x \in S^m.$$

Now employing the resolution method of Yager [87] for the double fuzzy constraints (as discussed here in Section 7.2) and following Zimmermann's approach [90], the above pair of fuzzy linear programming problems reduces to

(*GFP*5)

$$\begin{aligned} \max \quad & \lambda \\ \text{subject to,} \quad & \\ & x^T \tilde{A} y \,\textcircled{$\geq$}\, \tilde{v} - \tilde{p}(1-\lambda), \text{ for all } y \in S^n, \\ & x \in S^m, \\ & \lambda \in [0,1], \end{aligned}$$

and,

(*GFD*5)

$$\begin{aligned} \max \quad & \eta \\ \text{subject to,} \quad & \\ & x^T \tilde{A} y \,\textcircled{$\leq$}\, \tilde{w} + \tilde{q}(1-\eta), \text{ for all } x \in S^m, \\ & y \in S^n, \\ & \eta \in [0,1]. \end{aligned}$$

Next, we utilize the defuzzification function $F : N(R) \to R$ for the constraints of (*GFP*5) and (*GFD*5), to get

(*GFP*6)

$$\begin{aligned} \max \quad & \lambda \\ \text{subject to,} \quad & \\ F(x^T\tilde{A}y) \geq & \; F(\tilde{v}) - F(\tilde{p})(1-\lambda), \text{ for all } y \in S^n, \\ e^Tx &= 1, \\ \lambda &\leq 1, \\ x, \lambda &\geq 0, \end{aligned}$$

and,

(*GFD*6)

$$\begin{aligned} \max \quad & \eta \\ \text{subject to,} \quad & \\ F(x^T\tilde{A}y) \leq & \; F(\tilde{w}) + F(\tilde{q})(1-\eta), \text{ for all } x \in S^m, \\ e^Ty &= 1, \\ \eta &\leq 1, \\ y, \eta &\geq 0. \end{aligned}$$

As we have mentioned earlier, the defuzzification function F preserves the ranking when fuzzy numbers are multiplied by non-negative scalars, the problems (*GFP*6) and (*GFD*6) respectively become

(*GFP*7)

$$\begin{aligned} \max \quad & \lambda \\ \text{subject to,} \quad & \\ x^TF(\tilde{A})y \geq & \; F(\tilde{v}) - F(\tilde{p})(1-\lambda), \text{ for all } y \in S^n, \\ e^Tx &= 1, \\ \lambda &\leq 1, \\ x, \lambda &\geq 0, \end{aligned}$$

and,

(*GFD*7)

$$\begin{aligned} \max \quad & \eta \\ \text{subject to,} \quad & \\ x^TF(\tilde{A})y \leq & \; F(\tilde{w}) + F(\tilde{q})(1-\eta), \text{ for all } x \in S^m, \\ e^Ty &= 1, \\ \eta &\leq 1, \\ y, \eta &\geq 0, \end{aligned}$$

where $F(\tilde{A})$ is a (crisp) $(m \times n)$ matrix having entries as $F(\tilde{a}_{ij})$, $i = 1, 2, \ldots, m$ and $j = 1, 2, \ldots, n$.

Since S^m and S^n are convex polytopes, it is sufficient to consider only the extreme points (i.e. pure strategies) of S^m and S^n in the constraints of (*GFP*7) and (*GFD*7). This observation leads to the following two fuzzy linear programming problems, (*GFP*8) and (*GFP*8), for Player I and Player II respectively

(*GFP*8) max λ
subject to,

$$\begin{aligned} x^T F(\tilde{A})_j &\geq F(\tilde{v}) - F(\tilde{p})(1-\lambda), (j = 1, 2, \ldots, n), \\ e^T x &= 1, \\ \lambda &\leq 1, \\ x, \lambda &\geq 0, \end{aligned}$$

and,

(*GFD*8) max η
subject to,

$$\begin{aligned} F(\tilde{A})_i y &\leq F(\tilde{w}) + F(\tilde{q})(1-\eta), (i = 1, 2, \ldots, m), \\ e^T y &= 1, \\ \eta &\leq 1, \\ y, \eta &\geq 0. \end{aligned}$$

Here $F(\tilde{A})_i$ (respectively $F(\tilde{A})_j$) denotes the ith row (respectively j^{th} column) of $F(\tilde{A})$ where $i = 1, 2, \ldots, m$, and $j = 1, 2, \ldots, n$.

From the above discussion we now observe that for solving the fuzzy matrix game FG we have to solve the crisp linear programming problems (*GFP*8) and (*GFD*8) for Player I and Player II respectively. Also, if (x^*, λ^*) is an optimal solution of (*GFP*8) then x^* is an optimal strategy for Player I and λ^* is the degree to which the aspiration level $F(\tilde{v})$ of Player I can be met by choosing to play the strategy x^*. Similar interpretation can also be given to an optimal solution (y^*, η^*) of the problem (*GFD*8). All the results discussed in this section can now be summarized in the form of Theorem 7.6.1 given below.

Theorem 7.6.1 *The fuzzy matrix game FG described by* $FG = (S^m, S^n, \tilde{A}, \tilde{v}, \gtrsim, \tilde{p}, \tilde{w}, \lesssim, \tilde{q})$ *is equivalent to two crisp linear programming problems* (*GFP*8) *and* (*GFD*8) *which constitute a primal-dual pair in the sense of duality for linear programming in a fuzzy environment.*

Remark 7.6.1. It is important to note that the crisp problems (*GFP*8) and (*GFD*8) do not constitute a primal-dual pair in the conventional sense of duality in linear programming but are dual in "fuzzy" sense as explained above. Therefore if (x^*, λ^*) is optimal to (*GFP*8) and (y^*, η^*) is optimal to (*GFD*8) then in general one should not expect that $\lambda^* = \eta^*$.

Remark 7.6.2. If both players have the same aspiration level, i.e. $F(\tilde{v}) = F(\tilde{w})$ and in the optimal solutions of $(GFP8)$ and $(GFD8)$ $\lambda^* = \eta^* = 1$, then the fuzzy game FG reduces to the crisp two person zero sum game G. Thus for $F(\tilde{v}) = F(\tilde{w})$, $\lambda^* = \eta^* = 1$, FG reduces to G; the pair $(GFP8)$-$(GFD8)$ reduces to the crisp primal-dual pair (LP)-(LD).

Now the solution procedure for solving such a game FG is illustrated through the following numerical example:

Example 7.6.3. Consider the fuzzy game defined by the matrix of fuzzy numbers :

$$\tilde{\mathbf{A}} = \begin{bmatrix} \tilde{180} & \tilde{156} \\ \tilde{90} & \tilde{180} \end{bmatrix}$$

where $\tilde{180} = (175, 180, 190), \tilde{156} = (150, 156, 158)$ and $\tilde{90} = (80, 90, 100)$. Assuming that Player I and Player II have the tolerance levels $\tilde{p_1} = \tilde{p_2} = (14, 16, 20)$, and $\tilde{q_1} = \tilde{q_2} = (13, 15, 22)$. The aspiration levels for the Player I and Player II are $\tilde{v} = (155, 165, 175)$ and $\tilde{w} = (170, 180, 190)$ respectively.

According to Theorem 7.6.1, to solve this game we have to solve following two crisp linear programming problems $(GFP1)$ and $(GFD1)$ for Player I and Player II respectively:

$(GFP8)$ max λ

subject to,

$$\begin{aligned} 545x_1 + 270x_2 &\geq 495 - (1 - \lambda)(50) \\ 464x_1 + 545x_2 &\geq 495 - (1 - \lambda)(50), \\ x_1 + x_2 &= 1, \\ \lambda &\leq 1 \\ x_1, x_2, \lambda &\geq 0, \end{aligned}$$

and

$(GFD8)$ min $-\eta$

subject to,

$$\begin{aligned} 545y_1 + 464y_2 &\leq 540 + (1 - \eta)(50) \\ 270y_1 + 545y_2 &\leq 540 + (1 - \eta)(50) \\ y_1 + y_2 &= 1 \\ \eta &\leq 1 \\ y_1, y_2, \eta &\geq 0. \end{aligned}$$

Solving the above linear programming problems, we obtain ($x_1^* = 0.7725$, $x_2^* = 0.2275$, $\lambda^* = 0.5486$) and ($y_1^* = 0.0545$, $y_2^* = 0.9455$, $\eta^* = 1$). Therefore we obtain optimal strategies for Player I and Player II as ($x_1^* = 0.7725$, $x_2^* = 0.2275$) and ($y_1^* = 0.0545$, $y_2^* = 0.9455$) respectively and the two problems are dual to each other in the "fuzzy" sense as has been discussed earlier.

7.7 Conclusion

This chapter is mainly devoted to the study of matrix games with fuzzy pay-offs via a ranking (defuzzification) function approach. For this, first a suitable duality is established for linear programming problems with fuzzy parameters and then the same is employed to develop a solution procedure for solving two person zero sum matrix games with fuzzy pay-offs. The whole development presented here depends heavily on the resolution method of analyzing the system of double fuzzy inequalities of the type $\tilde{A}x \lesssim \tilde{b}$. In the literature, there are many other approaches for analyzing the system of double fuzzy inequalities of the type $\tilde{A}x \lesssim \tilde{b}$, the most notable being that of "modalities" due to Dubois and Prade [16] which has been extended directly to linear programming problems by Inuiguchi et al. ([26], [27], [28]). Recently, Inuiguchi et al. [29] advocated yet another approach for the system $\tilde{A}x \lesssim \tilde{b}$ which is based on fuzzy (valued) relations. In this chapter we have deliberately followed the Zimmermann type approach as discussed above for the system $\tilde{A}x \lesssim \tilde{b}$ so as to be in complete conformity with the notations and terminology of Campos [10]. Further, it may be noted that the ranking function approach as presented here to solve matrix games with both fuzzy goals as well as fuzzy pay-offs, is significantly different from that of Nishizaki and Sakawa [61]. In Nishizaki and Sakawa [61] the solution of such game is obtained by solving a somewhat difficult optimization problem in which the constraints are fractional, but here only two simple linear programming problems need to be solved. Of course there may be some difficulty in choosing an appropriate ranking function for the given fuzzy scenario. But once a "good" choice of ranking function has been made, then this approach seems to be attractive because of its simplicity. This discussion is further continued in Chapter 8 where some more models are presented for studying such fuzzy matrix games.

8

More on matrix games with fuzzy pay-offs

8.1 Introduction

This chapter is in continuation with the last chapter, and is devoted to the study of two more models of matrix games with fuzzy pay-offs. These models are Maeda's model [50] and Li's model [39]. The first model (Maeda [50]) is based on fuzzy max order to define three kinds of minmax equilibrium strategies so as to relate a (crisp) bi-matrix game with the given fuzzy matrix game, while the model due to Li [39] solves a matrix game with fuzzy pay-offs by solving a pair of related multi objective linear programming problems. These models seem to have some advantage over those prescribed in Chapter 7 because in certain situations it may be difficult to choose a suitable defuzzification (ranking) function.

This chapter has three main sections, namely, *definitions and preliminaries, a (crisp) bi-matrix game approach for matrix games with fuzzy pay-offs: Maeda's model,* and, *a two level linear programming approach for matrix games with fuzzy pay-offs: Li and Yang's model.*

8.2 Definitions and preliminaries

In this section, we recall some definitions and preliminaries introduced in earlier chapters and also introduce some new ones which are used here.

Definition 8.2.1 (Symmetric triangular fuzzy number). *Let* $m \in \mathbb{R}$ *and* $h > 0$. *Then a fuzzy number* $\tilde{a}$ *is called a symmetric triangular fuzzy number if its membership function is given by*

$$\mu_{\tilde{a}}(x) = \begin{cases} 1 - \left|\dfrac{x-m}{h}\right| & , x \in [m-h,\ m+h], \\ 0 & , otherwise \end{cases}$$

Here m is called the center and h is called the deviation parameter of $\tilde{a}$. A symmetric triangular fuzzy number will be denoted by $\tilde{a} = (m,\ h)_T$. Clearly a triangular fuzzy number $\tilde{a} = (a_l,\ a,\ a_u)$ will be symmetric if $a - a_l = a_u - a$ and in that case $m = a$ and $h = a - a_l = a_u - a$, i.e. $\tilde{a} = (m-h,\ m,\ m+h) = (m,\ h)_T$.

Let us recall that for a fuzzy number $\tilde{a}$ its α-cuts are given by

$$A_\alpha = \{\, x \in \mathbb{R} :\ \mu_{\tilde{a}}(x) \geq \alpha \,\},\ \alpha \in (0,1].$$

In the context of fuzzy numbers sometimes it is convenient to denote A_α by $[\tilde{a}]^\alpha$, $\alpha \in (0,1]$. Since $\alpha = 0$ is not covered in this definition, we define A_0 (or equivalently $[\tilde{a}]^0$) separately as follows.

For $\alpha = 0$, the set $[\tilde{a}]^0$ is defined as the closure of the set $\{\, x \in \mathbb{R} :\ \mu_{\tilde{a}}(x) > 0 \,\}$. Thus for each $\alpha \in [0,1]$, the set $[\tilde{a}]^\alpha$ is a closed interval $[a^L_\alpha,\ a^R_\alpha]$ where $a^L_\alpha = \inf\, [\tilde{a}]^\alpha$ and $a^R_\alpha = \sup\, [\tilde{a}]^\alpha$

For defining various minmax equilibrium strategies we shall need to compare two given elements of $\mathbb{R}^n$. This comparison will be done in accordance with the following understanding (Mangasarian [53]).

For $x, y \in \mathbb{R}^n$, we shall write

(i) $x \geqq y$ if and only if $x_i \geqq y_i$, $(i = 1, 2 \ldots, n)$
(ii) $x \geq y$ if and only if $x \geqq y$, and, $x \neq y$, and,
(iii) $x > y$ if and only if $x_i > y_i$, $(i = 1, 2 \ldots, n)$

These notations are normally used in the context of multiobjective optimization only and, unless otherwise stated, we shall restrict their usage in this chapter only.

Definition 8.2.2 (Ordering between fuzzy numbers). *Let $\tilde{a}$ and $\tilde{b}$ two fuzzy numbers. Then treating $(a^L_\alpha,\ a^R_\alpha)$ and $(b^L_\alpha,\ b^R_\alpha)$ as vectors in $\mathbb{R}^2$ and following Mangasarian's [53] notations, we define the fuzzy max order '$\succeqq$', the strict fuzzy max order '$\succeq$' and strong fuzzy max order '$\succ$' as follows*

(i) $\tilde{a} \succeqq \tilde{b}$ if $(a^L_\alpha,\ a^R_\alpha)^T \geqq (b^L_\alpha,\ b^R_\alpha)^T$, for all $\alpha \in [0,1]$,
(ii) $\tilde{a} \succeq \tilde{b}$ if $(a^L_\alpha,\ a^R_\alpha)^T \geq (b^L_\alpha,\ b^R_\alpha)^T$, for all $\alpha \in [0,1]$,
(iii) $\tilde{a} \succ \tilde{b}$ if $(a^L_\alpha,\ a^R_\alpha)^T > (b^L_\alpha,\ b^R_\alpha)^T$, for all $\alpha \in [0,1]$.

The following theorem gives a characterization of above orders for the case of symmetric triangular fuzzy numbers.

Theorem 8.2.1 (Furukawa [19]). *Let $\tilde{a} = (a, \alpha)_T$ and $\tilde{b} = (b, \beta)_T$ be two symmetric triangular fuzzy numbers. Then*

(i) $\tilde{a} \geqq \tilde{b} \Longleftrightarrow |\alpha - \beta| \leqq (a - b)$
(ii) $\tilde{a} \succ \tilde{b} \Longleftrightarrow |\alpha - \beta| < (a - b)$

8.3 A bi-matrix game approach: Maeda's model

Consider a two person zero-sum matrix game with fuzzy pay-offs described by $FG = (S^m, S^n, \tilde{A})$, where the symbols have their usual meanings as explained in Chapter 7. We now have the following three types of concepts of equilibrium strategies for the game FG (Maeda [50]).

Definition 8.3.1 (Minmax equilibrium strategy). *An element $(\bar{x}, \bar{y}) \in S^m \times S^n$ is called a minmax equilibrium strategy of the game $FG = (S^m, S^n, \tilde{A})$ if*

(i) $x^T\tilde{A}\bar{y} \leqq \bar{x}^T\tilde{A}\bar{y}$, *for all* $x \in S^m$, *and*
(ii) $\bar{x}^T\tilde{A}\bar{y} \leqq \bar{x}^T\tilde{A}y$, *for all* $y \in S^n$.

In this case the scalar $\tilde{v} = \bar{x}^T\tilde{A}\bar{y}$ is said to be the (fuzzy) value of the game FG and the triplet $(\bar{x}, \bar{y}, \tilde{v})$ is said to be a solution of FG under the fuzzy max order '$\geqq$'.

Definition 8.3.2 (Non-dominated minmax equilibrium strategy). *An element $(\bar{x}, \bar{y}) \in S^m \times S^n$ is said to be a non-dominated minmax equilibrium strategy of the game $FG = (S^m, S^n, \tilde{A})$ if*

(i) there does not exist any $x \in S^m$ such that $\bar{x}^T\tilde{A}\bar{y} \leq x^T\tilde{A}\bar{y}$, and
(ii) there does not exist any $y \in S^n$ such that $\bar{x}^T\tilde{A}y \leq \bar{x}^T\tilde{A}\bar{y}$.

Definition 8.3.3 (Weak non-dominated minmax equilibrium strategy.) *An element $(\bar{x}, \bar{y}) \in S^m \times S^n$ is said to be a weak non-dominated minmax equilibrium strategy of the game $FG = (S^m, S^n, \tilde{A})$ if*

(i) there does not exist any $x \in S^m$ such that $\bar{x}^T\tilde{A}\bar{y} \prec x^T\tilde{A}\bar{y}$, and
(ii) there does not exist any $y \in S^n$ such that $\bar{x}^T\tilde{A}y \prec \bar{x}^T\tilde{A}\bar{y}$.

Remark 8.3.1. Minmax equilibrium strategy $\Rightarrow$ non-dominated minmax equilibrium strategy $\Rightarrow$ weak non-dominated minmax equilibrium strategy.

Remark 8.3.2. If $\tilde{A}$ is crisp, i.e. the game *FG* is the usual two person zero sum matrix game $G = (S^m, S^n, A)$, then all the above three definitions coincide and become the definition of the usual saddle point.

We now assume that the elements $\tilde{a}_{ij}$ of the pay-off matrix $\tilde{A} = (\tilde{a}_{ij})$ are symmetric triangular fuzzy numbers given by $\tilde{a}_{ij} = (a_{ij}, h_{ij})_T$ and $A = (a_{ij})$ and $H = (h_{ij})$ are two $(m \times n)$ crisp matrices resulting from the fuzzy matrix $\tilde{A}$.

If we now agree to denote by $[\tilde{A}]^\alpha = ([\tilde{a}_{ij}]^\alpha)$ where $[\tilde{a}_{ij}]^\alpha$ is the α-level set of the fuzzy number $\tilde{a}_{ij}$ then for the symmetric TFN case

$$[\tilde{A}]^\alpha = \left(A - (1-\alpha)H,\ A + (1-\alpha)H\right).$$

Here it may be noted that $[\tilde{A}]^\alpha$ is not a matrix in the conventional sense but rather it is an arrangement of $(m \times n)$ intervals of the type $\left(a_{ij} - (1-\alpha)\, h_{ij},\ a_{ij} + (1-\alpha)\, h_{ij}\right)$, $(i = 1,2\ldots,m,\ j = 1,2\ldots,n)$.

Further, let fuzzy matrix games with symmetric TFN's pay-offs be denoted by $SFG = (S^m, S^n, \tilde{A})$.

The following theorem now gives a characterization of the minmax equilibrium strategy of the game $FG = (S^m, S^n, \tilde{A})$ for the case of symmetric TFN's, i.e. for the game *SFG*.

Theorem 8.3.1 *A point $(\bar{x}, \bar{y}) \in S^m \times S^n$ is a minmax equilibrium strategy of the game $SFG = (S^m, S^n, \tilde{A})$, $\tilde{A} = (\tilde{a}_{ij})$ with $\tilde{a}_{ij} = (a_{ij}, h_{ij})_T$, if and only if*

(i) $x^T(A+H)\bar{y} \leqq \bar{x}^T(A+H)\bar{y} \leqq \bar{x}^T(A+H)y$, *and*
(ii) $x^T(A-H)\bar{y} \leqq \bar{x}^T(A-H)\bar{y} \leqq \bar{x}^T(A-H)y$,

hold for all $x \in S^m$, $y \in S^n$.

Proof. Let $(\bar{x}, \bar{y}) \in S^m \times S^n$ be a minmax equilibrium strategy of the game *SFG*. Then from Theorem 8.2.1 we have

$$|\, \bar{x}^T H\bar{y} - x^T H\bar{y} \,| \leqq \bar{x}^T A\bar{y} - x^T A\bar{y},$$

and

$$|\, \bar{x}^T H y - \bar{x}^T H\bar{y} \,| \leqq \bar{x}^T A y - \bar{x}^T A\bar{y},$$

for all $x \in S^m, y \in S^n$.
The result now follows directly by employing the definition of the modulus function and appropriate rearrangement of terms.

Remark 8.3.3. In view of the above theorem if we wish to solve the game SFG, then we have to consider a pair of crisp two person zero-sum games $G_1 = (S^m, S^n, A+H)$ and $G_2 = (S^m, S^n, A-H)$, and attempt to determine a point $(\bar{x}, \bar{y}) \in S^m \times S^n$ which is simultaneously a saddle point of G_1 and G_2. Since this is not going to happen in general, the next best thing will be to look for solution in accordance with the non-dominated minmax equilibrium strategy (Definition 8.3.2) or weak non-dominated minmax equilibrium strategy (Definition 8.3.3). The development given below asserts that the game *SFG* certainly has a solution in these situations.

Theorem 8.3.2 *Let $x^T \mathcal{A} y$ be defined as the ordered pair*

$$\left(x^T(A+H)y,\ x^T(A-H)y\right)\ \forall x \in S^m,\ y \in S^n.$$

Then an element $(\bar{x}, \bar{y}) \in S^m \times S^n$ is a non-dominated minmax equilibrium strategy of the game SFG if and only if

(i) there does not exist any $x \in S^m$ such that $\bar{x}^T \mathcal{A} \bar{y} \leq x^T \mathcal{A} \bar{y}$, and
(ii) there does not exist any $y \in S^n$ such that $\bar{x}^T \mathcal{A} y \leq \bar{x}^T \mathcal{A} \bar{y}$.

Proof. We shall first prove the direct part. For this let us assume that $(\bar{x}, \bar{y}) \in S^m \times S^n$ is a non-dominated minmax equilibrium strategy of the game *SFG*. If possible, let there exist $\hat{x} \in S^m$ such that $\bar{x}^T \mathcal{A} \bar{y} \leq (\hat{x})^T \mathcal{A} \bar{y}$. This, by definition, implies

$$\left(\bar{x}^T(A-H)\bar{y},\ \bar{x}^T(A+H)\bar{y}\right)^T \leq \left(\hat{x}^T(A-H)\bar{y},\ \hat{x}^T(A+H)\bar{y}\right)^T.$$

Now by appropriately rearranging terms in the above inequality we get

$$\left(\bar{x}^T A\bar{y} - \hat{x}^T A\bar{y},\ \bar{x}^T A\bar{y} - \hat{x}^T A\bar{y}\right)^T \leq \left(\bar{x}^T H\bar{y} - \hat{x}^T H\bar{y},\ \hat{x}^T H\bar{y} - \bar{x}^T H\bar{y}\right)^T.$$

As both components on the L.H.S are same but on the R.H.S, they are negative of the other, we have

$$\bar{x}^T A\bar{y} - (\hat{x})^T A\bar{y} < 0,$$

i.e

$$\bar{x}^T A\bar{y} < \hat{x}^T A\bar{y}.$$

Also in view of the specific understanding of the symbol '$\leq$', the above inequality with some obvious manipulations, gives that for all $\alpha \in [0,1]$,

$$\left(\bar{x}^T(A-(1-\alpha)H)\bar{y},\ \bar{x}^T(A+(1-\alpha)H)\bar{y}\right)^T \leq \left(\hat{x}^T(A-(1-\alpha)H)\bar{y},\ \hat{x}^T(A+(1-\alpha)H)\bar{y}\right)^T,$$

which by Definition 8.2.2 implies that $\bar{x}^T\tilde{A}y \preceq \hat{x}^T\tilde{A}\bar{y}$. But this is a contradiction to the fact that $(\bar{x},\ \bar{y})$ is a non-dominated minmax equilibrium strategy. Similarly we can show that there does not exist any $y \in S^n$ such that $\bar{x}^T\mathcal{A}y \leq \bar{x}^T\mathcal{A}\bar{y}$.
Conversely, let $(\bar{x},\ \bar{y}) \in S^m \times S^n$ be such that conditions (i) and (ii) of above theorem hold. We have to show that $(\bar{x},\ \bar{y})$ is a non-dominated minmax equilibrium strategy of the game *SFG*.
For this, if possible, let there exist $\hat{x} \in S^m$ such that $\bar{x}^T\tilde{A}\bar{y} \preceq \hat{x}^T\tilde{A}\bar{y}$. This, by definition, gives

$$\left(\bar{x}^T(A-H)\bar{y},\ \bar{x}^T(A+H)\bar{y}\right)^T \leq \left(\hat{x}^T(A-H)\bar{y},\ \hat{x}^T(A+H)\bar{y}\right)^T,$$

which implies that

$$\bar{x}^T\mathcal{A}\bar{y} \leq (\hat{x})^T\mathcal{A}\bar{y},$$

but this is a contradiction to the condition (i) of the theorem. Similarly we can show that if there exists $\hat{y} \in S^n$ such that $\bar{x}^T\tilde{A}\hat{y} \leq (\bar{x})^T\tilde{A}\bar{y}$ then condition (ii) of the theorem is contradicted.

Theorem 8.3.3 *An element $(\bar{x},\ \bar{y}) \in S^m \times S^n$ is a weak non-dominated minmax equilibrium strategy of the game SFG if and only if*

(i) there does not exist any $x \in S^m$ such that $\bar{x}^T\mathcal{A}\bar{y} < x^T\mathcal{A}\bar{y}$, and
(ii) there does not exist any $y \in S^n$ such that $\bar{x}^T\mathcal{A}y < \bar{x}^T\mathcal{A}\bar{y}$.

The proof of Theorem 8.3.3 is similar to that of Theorem 8.3.2. In view of Theorems 8.3.1-8.3.3, it is natural to define the bi-matrix game $BG(\lambda,\mu) = \left(S^m,\ S^n,\ A(\lambda),\ -A(\mu)\right)$ for $\lambda,\ \mu \in [0,1]$, where $A(\lambda) = A + (1-2\lambda)H$, and $A(\mu) = A + (1-2\mu)H$. Here it may be noted that $A(0) = A + H$, $A(1) = A - H$, and for $\lambda = \mu$, $BG(\lambda,\mu)$ becomes the matrix game $\left(S^m,\ S^n,\ A(\lambda)\right)$.

We now recall the following from Chapter 1.

Definition 8.3.4 (Nash equilibrium strategy). *Let $\lambda,\ \mu \in [0,1]$. A point $(\bar{x},\ \bar{y}) \in S^m \times S^n$ is called a Nash equilibrium strategy of the game $BG(\lambda,\mu)$ if*

(i) $x^TA(\lambda)\bar{y} \leqq \bar{x}^TA(\lambda)\bar{y}$, for all $x \in S^m$, and
(ii) $\bar{x}^TA(\mu)\bar{y} \leqq \bar{x}^TA(\mu)y$, for all $y \in S^n$.

Theorem 8.3.4 *An element $(\bar{x},\ \bar{y}) \in S^m \times S^n$ is a non-dominated minmax equilibrium strategy of the game SFG if and only if there exist* *positive real numbers λ, $\mu \in (0,1)$ such that $(\bar{x}, \bar{y})$ is a Nash equilibrium strategy of the (crisp) bi-matrix game $BG(\lambda, \mu)$.*

Proof. Let $(\bar{x}, \bar{y}) \in S^m \times S^n$ be a non-dominated minmax equilibrium strategy of the game *SFG*. Therefore, by Theorem 8.3.2, there does not exist any $x \in S^m$ such that $\bar{x}^T \mathcal{A} \bar{y} \leq x^T \mathcal{A} \bar{y}$ i.e. there does not exist any $x \in S^m$ such that

$$\left(\bar{x}^T(A-H)\bar{y},\ \bar{x}^T(A+H)\bar{y}\right)^T \leq \left(x^T(A-H)\bar{y},\ x^T(A+H)\bar{y}\right)^T.$$

But this implies (Steuer [71]) that there exist positive scalars λ_1, λ_2 with $\lambda_1 + \lambda_2 = 1$ such that for all $x \in S^m$,

$$\bar{x}^T\Big(\lambda_1(A+H) + \lambda_2(A-H)\Big)\bar{y} \geqq x^T\Big(\lambda_1(A+H) + \lambda_2(A-H)\Big)\bar{y}.$$

Now by taking $\lambda_2 = \lambda$ and $\lambda_1 = (1-\lambda)$, the above inequality becomes

$$\bar{x}^T\Big(A + (1-2\lambda)H\Big)\bar{y} \geqq x^T\Big(A + (1-2\lambda)H\Big)\bar{y}, \text{ for all } x \in S^m.$$

Similarly the second condition of Theorem 8.3.2 gives (Steuer [71]) that there exists $0 < \mu < 1$ such that

$$\bar{x}^T\Big(A + (1-2\mu)H\Big)\bar{y} \leqq \bar{x}^T\Big(A + (1-2\mu)H\Big)y, \text{ for all } y \in S^n.$$

The above two inequalities therefore imply that $(\bar{x}, \bar{y})$ is a Nash equilibrium strategy of the bi-matrix game $BG(\lambda, \mu)$.
Conversely let λ, $\mu \in (0,1)$ and $(\bar{x}, \bar{y})$ be a Nash equilibrium strategy of the game $BG(\lambda, \mu)$. Then by Definition 8.3.4

$$\bar{x}^T\Big(A + (1-2\lambda)H\Big)\bar{y} \geqq x^T\Big((A + (1-2\lambda)H\Big)\bar{y}, \text{ for all } x \in S^m,$$

and

$$\bar{x}^T\Big(A + (1-2\mu)H\Big)y \geqq \bar{x}^T\Big(A + (1-2\mu)H\Big)\bar{y}, \text{ for all } y \in S^n.$$

But

$$A + (1-2\lambda)H = \lambda(A-H) + (1-\lambda)(A+H)$$

and

$$A + (1-2\mu)H = \mu(A-H) + (1-\mu)(A+H).$$

Therefore the above inequalities imply that for λ, $\mu \in (0,1)$,

$$\bar{x}^T\Big(\lambda(A-H) + (1-\lambda)(A+H)\Big)\bar{y} \geqq x^T\Big(\lambda(A-H) + (1-\lambda)(A+H)\Big)\bar{y},\ x \in S^m,$$

and

$$\bar{x}^T\Big(\mu(A-H)+(1-\mu)(A+H)\Big)y \geqq \bar{x}^T\Big(\mu(A-H)+(1-\mu)(A+H)\Big)\bar{y},\ y \in S^n,$$

which again means (Steuer [71]) that there is no $x \in S^m$ such that

$$\left(\bar{x}^T(A-H)\bar{y},\ \bar{x}^T(A+H)\bar{y}\right)^T \leq \left(x^T(A-H)\bar{y},\ x^T(A+H)\bar{y}\right)^T,$$

and also there is no $y \in S^n$ such that

$$\left(\bar{x}^T(A-H)y,\ \bar{x}^T(A+H)y\right)^T \leq \left(\bar{x}^T(A-H)\bar{y},\ \bar{x}^T(A+H)\bar{y}\right)^T.$$

The above inequalities now imply that $(\bar{x},\ \bar{y})$ is a non-dominated minmax equilibrium strategy of the game SFG.

Theorem 8.3.5 *An element $(\bar{x},\ \bar{y}) \in S^m \times S^n$ is a weak non-dominated minmax equilibrium strategy, of the fuzzy matrix game SFG if and only if there exist real numbers $\lambda,\ \mu \in [0,1]$ such that $(\bar{x},\ \bar{y})$ is a Nash equilibrium strategy of the bi-matrix game $BG(\lambda,\mu)$.*

The proof of the above theorem is similar to that of Theorem 8.3.4.

Theorem 8.3.6 *For the fuzzy matrix game SFG the following is true*

(i) there exists at least one non-dominated minmax equilibrium strategy,
(ii) there exists at least one weak non-dominated minmax equilibrium strategy.

The proof of Theorem 8.3.6 follows because the existence of a Nash equilibrium strategy for the (crisp) bi-matrix $BG(\lambda,\mu)$ is always guaranteed as stated in Theorem 1.5.1.

In view of Theorems 8.3.4 and 8.3.5, for finding a non-dominated (weak non-dominated) minmax equilibrium strategy of the fuzzy matrix game SFG we have to find Nash equilibrium strategies of the (crisp) bi-matrix game $BG(\lambda,\mu)$.

Example 8.3.4. (Maeda [50]). Let us consider the fuzzy matrix game SFG where the pay off matrix $\tilde{A}$ is given by

$$\tilde{A} = \begin{pmatrix} (180,5)_T & (156,6)_T \\ (90,10)_T & (180,5)_T \end{pmatrix}.$$

Now as explained in Theorem 8.3.4, to solve the fuzzy matrix game SFG, we have to consider the (crisp) bi- matrix game $BG(\lambda,\mu) =$

$\left(S^m,\ S^n,\ A(\lambda),\ -A(\mu)\right)$, where $\lambda,\ \mu \in (0,1)$ and $A(\lambda) = A + (1-2\lambda)H$ and $A(\mu) = A + (1-2\mu)H$. For the given fuzzy matrix $\tilde{A}$, we have

$$A(\lambda) = \begin{pmatrix} 185-10\lambda & 162-12\lambda \\ 100-20\lambda & 185-10\lambda \end{pmatrix},$$

and

$$A(\mu) = \begin{pmatrix} 185-10\mu & 162-12\mu \\ 100-20\mu & 185-10\mu \end{pmatrix}.$$

Also for the (crisp) bi-matrix game $BG = \left(S^m,\ S^n,\ A(\lambda),\ -A(\mu)\right)$, an element $(x^*,\ y^*) \in S^2 \times S^2$ with $x^* = (x_1^*,\ x_2^*) = \left(\dfrac{85+10\mu}{108+12\mu}, \dfrac{23+2\mu}{108+12\mu}\right)$, $y^* = (y_1^*,\ y_2^*) = \left(\dfrac{23+2\lambda}{108+12\lambda}, \dfrac{85+10\lambda}{108+12\lambda}\right)$, gives a Nash equilibrium strategy of the game $BG(\lambda,\mu) = \left(S^2,\ S^2,\ A(\lambda),\ -A(\mu)\right)$.
Therefore, from Theorem 8.3.4 the set NDM of all non-dominated minmax equilibrium strategies of the fuzzy matrix game SFG is given by

$$NDM = \{(x^*,\ y^*)\} = \left\{ \left((x_1^*,\ x_2^*),\ (y_1^*, y_2^*)\right) :\ \lambda,\ \mu \in (0,1) \right\},$$

where x^* and y^* are as determined earlier.
In a similar manner, from Theorem 8.3.5, the set $WNDM$ of all weak non-dominated minmax equilibrium strategies is given by

$$WNDM = \left\{ \left((x_1^*,\ x_2^*),\ (y_1^*,\ y_2^*)\right) :\ \lambda,\ \mu \in [0,1] \right\}.$$

Remark 8.3.5. For $\lambda = \mu$, the bi-matrix game $BG(\lambda,\mu)$ of the above example becomes the two person zero sum game $G = \left(S^2,\ S^2,\ A(\lambda)\right)$, $\lambda \in (0,1)$ and $x^* = (x_1^*, x_2^*)$, $y^* = (y_1^*,\ y_2^*)$ become optimal strategies for Player I and II respectively. As shown by Maeda [50], this result is the same as that of Campos [10]. However as the basic idea in Campos [10], is to convert a fuzzy matrix game to matrix game with crisp pay-offs, this approach can not be used for the case $\lambda \neq \mu$. In this sense, Maeda's [50] approach is different from Campos [10] and is more general.

8.4 A multiobjective programming approach: Li's model

In this section we present Li's model [39] for solving the fuzzy matrix game $FG = (S^m,\ S^n,\ \tilde{A})$, where $\tilde{A} = (\tilde{a}_{ij})$ with $\tilde{a}_{ij}$ being a TFN. The approach taken by Li [39] is different from Maeda [50] as it uses a different "ordering" for fuzzy numbers and also constructs a pair of (crisp) multiobjective linear programming problems for the given fuzzy game FG. These multiobjective linear programming problems are then *solved* to obtain a "*solution*" of the given fuzzy matrix game FG. As it happens, the solution procedure of Li [39] does not provide the complete solution of the game FG. However Li and Yang [42] very recently proposed a new two level linear programming approach to solve these multiobjective linear programming problems so as to provide a complete solution of the fuzzy matrix game FG. The presentation described below is based on Li [39] and Li and Yang [42].

Definition 8.4.1 (Ordering of TFN's). *Let* $\tilde{a} = (a_l,\ a,\ a_u)$ *and* $\tilde{b} = (b_l,\ b,\ b_u)$ *be two TFN's. Then* $\tilde{a} \lesssim \tilde{b}$ *if* $a_l \leqq b_l,\ a \leqq b,\ a_u \leqq b_u$. *The symbol* $\tilde{a} \gtrsim \tilde{b}$ *is defined similarly.*

Remark 8.4.1. The ordering of triangular fuzzy numbers as described above is a special case of ordering of general fuzzy numbers as introduced by Ramik and Rimanek [65]. Specifically, given two fuzzy numbers $\tilde{a}$ and $\tilde{b}$ in accordance with Ramik and Rimanek [66], $\tilde{a} \lesssim \tilde{b}$ if for $\alpha \in [0,1]$,

(i) $\sup\ [\tilde{a}]^\alpha \leqq \sup\ [\tilde{b}]^\alpha$, and,
(ii) $\inf\ [\tilde{a}]^\alpha \leqq \inf\ [\tilde{b}]^\alpha$.

Here $[\tilde{a}]^\alpha$, $\alpha \in [0,1]$, is the α-cut of the fuzzy number $\tilde{a}$ as described in Section 8.2. For the case where $\tilde{a}$ and $\tilde{b}$ are TFN's (following Ramik and Rimanek's [66] notations $\tilde{a} = (m,\ \alpha,\ \beta)$, $\tilde{b} = (n,\ \nu,\ \delta)$) this definition reduces to $m \leqq n,\ \nu - \alpha \leqq n - m,\ \beta - \delta \leqq n - m$, i.e. $m \leqq n,\ m - \alpha \leqq n - \nu,\ m + \beta \leqq n + \delta$. In our notation, $\tilde{a} = (a_l,\ a,\ a_u)$, i.e. $a = m,\ a_l = m - \alpha,\ a_u = m + \beta$ and similarly for $\tilde{b} = (b_l,\ b,\ b_u)$, we have i.e. $b = n,\ b_l = n - \nu,\ b_u = n + \delta$. Therefore for the specific case of TFN's, the definition of ordering of Ramik and Rimanek [66] becomes Definition 8.4.1.

We shall now introduce the concept of the solution of the fuzzy matrix game $FG = (S^m,\ S^n,\ \tilde{A})$ under the chosen ordering of TFN's. Here $\tilde{A} = (\tilde{a}_{ij})$, and $\tilde{a}_{ij} = ((\tilde{a}_{ij})_l,\ \tilde{a}_{ij},\ (\tilde{a}_{ij})_u)$ $(i = 1,2,\dots,m,\ j = 1,2,\dots,n)$ are TFN's.

Definition 8.4.2 (Reasonable solution of FG). *Let* $\tilde{v} = (v_l,\ v,\ v_u)$ *and* $\tilde{w} = (w_l,\ w,\ w_u)$ *be TFN's. Then* $(\tilde{v},\ \tilde{w})$ *is called a reasonable solution of the fuzzy matrix fame FG if there exist* $\bar{x} \in S^m$, $\bar{y} \in S^n$ *such that*

(i) $\bar{x}^T\tilde{A}y \tilde{\geq} \tilde{v}$ *for all* $y \in S^n$, *and*
(ii) $x^T\tilde{A}\bar{y} \tilde{\leq} \tilde{w}$ *for all* $x \in S^m$.

If $(\tilde{v}, \tilde{w})$ *is a reasonable solution of FG then* $\tilde{v}$ *(respectively* $\tilde{w}$*) is called the reasonable value of Player I (respectively Player II).*

Let V (respectively W) be the set of all reasonable values $\tilde{v}$ (respectively $\tilde{w}$) for Player I (respectively Player II). Then we have the following definition.

Definition 8.4.3 (Solution of the game ***FG*****).** *An element* $(\tilde{v}^*, \tilde{w}^*) \in V \times W$ *is called a solution of the game FG if*

(i) $\tilde{v}^* \tilde{\geq} \tilde{v}$ *for all* $\tilde{v} \in V$, *and*
(ii) $\tilde{w}^* \tilde{\leq} \tilde{w}$ *for all* $\tilde{w} \in W$.

In fact, a much better way will be to call $(x^*, y^*, \tilde{v}^*, \tilde{w}^*)$ as a solution of the game *FG*, where $x^* \in S^m$, $y^* \in S^n$ are strategies for which $(\tilde{v}^*, \tilde{w}^*)$ is a reasonable solution of the game *FG*. In that case x^* (respectively y^*) will be called an optimal strategy for Player I (respectively Player II), and $\tilde{v}^*$ (respectively $\tilde{w}^*$) the value of the game for Player I (respectively Player II).

In view of Definitions 8.4.2 and 8.4.3, to solve the fuzzy matrix game *FG* we should solve the following fuzzy optimization problems (*FOP*1) and (*FOP*2) for Player I and Player II respectively

(*FOP*1) max $\tilde{v}$
subject to,

$$x^T\tilde{A}y \tilde{\geq} \tilde{v}, \quad \text{for all } y \in S^n,$$
$$x \in S^m,$$

and

(*FOP*2) min $\tilde{w}$
subject to,

$$x^T\tilde{A}y \tilde{\leq} \tilde{w}, \quad \text{for all } x \in S^m,$$
$$y \in S^n.$$

Since $x \in S^m, y \in S^n$ and fuzzy inequalities '$\tilde{\geq}$' and '$\tilde{\leq}$' are preserved under positive multiplication, it makes sense to consider only extreme points of sets S^m and S^n in problems (*FOP*1) and (*FOP*2). This leads to problems

(*FOP*3) max $\tilde{v}$
subject to,

$$\sum_{i=1}^{m} \tilde{a}_{ij} x_i \gtrsim \tilde{v}, \ (j = 1, 2, \ldots, n),$$
$$e^T x = 1,$$
$$x \geq 0,$$

and

(*FOP*4) min $\tilde{w}$

subject to,

$$\sum_{j=1}^{n} \tilde{a}_{ij} y_j \lesssim w, \ (i = 1, 2, \ldots, m),$$
$$e^T y = 1,$$
$$y \geq 0.$$

As $\tilde{a}_{ij}$ $(i = 1, 2, \ldots, m \ j = 1, 2, \ldots, n)$ are TFN's, so should be $\tilde{v}$ and $\tilde{w}$, because only then the constraints of (*FOP*1) and (*FOP*2) (i.e. of (*FOP*3) and (*FOP*4)) will be meaningful. Let $\tilde{v} = (v_l, \ v, \ v_u)$, and $\tilde{w} = (w_l, \ w, \ w_u)$ be TFN's then problems (*FOP*3) and (*FOP*4) can respectively be rewritten as

(*FOP*5) max $(v_l, \ v, \ v_u)$

subject to,

$$\sum_{i=1}^{m} (a_{ij})_l x_i \geqq \ v_l, \ (j = 1, 2, \ldots, n),$$
$$\sum_{i=1}^{m} a_{ij} x_i \geqq \ v, \ (j = 1, 2, \ldots, n),$$
$$\sum_{i=1}^{m} (a_{ij})_u x_i \geqq \ v_u, \ (j = 1, 2, \ldots, n),$$
$$e^T x = 1,$$
$$x \geq 0,$$

and

(*FOP*6) min $(w_l, \ w, \ w_u)$

subject to,

$$\sum_{j=1}^{n}(a_{ij})_l y_j \leqq w_l, \quad (i=1,2,\ldots,m),$$
$$\sum_{j=1}^{n} a_{ij} y_j \leqq w, \quad (i=1,2,\ldots,m),$$
$$\sum_{j=1}^{n}(a_{ij})_u y_j \leqq w_u, \; (i=1,2,\ldots,m),$$
$$e^T y = 1,$$
$$y \geq 0.$$

Now (*FOP*5) and (*FOP*6) are (crisp) multiobjective linear programming problems. The main question now is that in what sense we should define the "solution" of (*FOP*5) (and (*FOP*6)) so that the "solution" so obtained is consistent with Definitions 8.4.1 and 8.4.3. It is not difficult to see that this will happen for example, if we say that (x^*, v^*) is optimal to (*FOP*5) provided $v_l^* \geqq v_l$, $v^* \geqq v$, $v_u^* \geqq v_u$ for all (x, v) feasible to (*FOP*5), where $v^* = (v_l^*, v^*, v_u^*)$ and $v = (v_l, v, v_u)$. This means that if we denote the set of all feasible solutions of (*FOP*5) by T then the three scalar optimization problems, namely $\max_T v_l$, $\max_T v$, $\max_T v_u$ achieve their optimal value for the same (x^*, v^*). This is something which is going to happen very rarely. Similar arguments hold for problem (*FOP*6) as well. Therefore probably the very definition of the solution of the game *FG* (Definition 8.4.3) should be modified.

Since the multiobjective programming problems are most satisfactorily discussed for the case of *Pareto optimal solutions*, we should define the "solutions of the game *FG*" in this sense only. This leads to the following definition.

Definition 8.4.4 (Solution of the game *FG*). *An element* $\left(\tilde{v}^* = (v_l^*, v^*, v_u^*),\ \tilde{w}^* = (w_l^*, w^*, w_u^*)\right) \in V \times W$ *is called a solution of the game FG if*

(i) there does not exist any $\tilde{v} = (v_l, v, v_u) \in V$ *such that* $(v_l, v, v_u) \geq (v_l^*, v^*, v_u^*)$, *and*
(ii) there does not exist any $\tilde{w} = (w_l, w, w_u) \in V$ *such that* $(w_l, w, w_u) \leq (w_l^*, w^*, w_u^*)$.

Here the orderings '$\geq$' and '$\leq$' in $\mathbb{R}^n$ are to be understood in the sense as discussed in Section 8.2 (Magasarian [53]).

Now to be in conformity with the above definition of the solution of the fuzzy matrix game *FG*, we should therefore take the multiobjective linear programming problems (*FOP*5) and (*FOP*6) and attempt to obtain their Pareto optimal solutions $\left(x^*,\ \tilde{v}^* = (v_l^*,\ v^*,\ v_u^*)\right)$ and $\left(y^*,\ \tilde{w}^* = (w_l^*,\ w^*,\ w_u^*)\right)$. These solutions will give (x^*, y^*, v^*, w^*) which could then be taken as "a solution of the fuzzy matrix game *FG*" in the sense of Definition 8.4.4. As problems (*FOP*5) and (*FOP*6) are (crisp) multiobjective linear programming problems it is better to denote them (*MOP*1) and (*MOP*2). Thus (*MOP*1) and (*MOP*2) are

$$
\begin{aligned}
(MOP1)\qquad & \max \quad (v_l,\ v,\ v_u)\\
& \text{subject to,}\\
& \sum_{i=1}^{m} (a_{ij})_l x_i \geqq v_l,\ (j = 1, 2, \ldots, n),\\
& \sum_{i=1}^{m} a_{ij} x_i \geqq v,\ \ (j = 1, 2, \ldots, n),\\
& \sum_{i=1}^{m} (a_{ij})_u x_i \geqq v_u,\ (j = 1, 2, \ldots, n),\\
& e^T x = 1,\\
& x \geq 0,
\end{aligned}
$$

and

$$
\begin{aligned}
(MOP2)\qquad & \min \quad (w_l,\ w,\ w_u)\\
& \text{subject to,}\\
& \sum_{j=1}^{n} (a_{ij})_l y_j \leqq w_l,\ (i = 1, 2, \ldots, m),\\
& \sum_{j=1}^{n} a_{ij} y_j \leqq w,\ \ (i = 1, 2, \ldots, m),\\
& \sum_{j=1}^{n} (a_{ij})_u y_j \leqq w_u,\ (i = 1, 2, \ldots, m),\\
& e^T y = 1,\\
& y \geq 0.
\end{aligned}
$$

respectively.

Li and Yang [42] suggested a two level linear programming approach to find solutions of (*MOP*1) and (*MOP*2) in the sense of Pareto optimality. We below discuss this approach for solving (*MOP*1); the details of solving (*MOP*2) will be similar and can be described on the same lines.

Level 1 : Consider the following scalar linear programming problem, namely Level-1: *LPP*,

$$\begin{aligned} \max \quad & v \\ \text{subject to,} \quad & \\ & \sum_{i=1}^{m} (a_{ij})_l x_i \geqq v_l, \ (j = 1,2,\ldots,n), \\ & \sum_{i=1}^{m} a_{ij} x_i \geqq v, \ (j = 1,2,\ldots,n), \\ & \sum_{i=1}^{m} (a_{ij})_u x_i \geqq v_u, \ (j = 1,2,\ldots,n), \\ & e^T x = 1, \\ & x \geq 0, \end{aligned}$$

Here the decision variables are $x = (x_1, x_2, \ldots x_n)$ and $(v_l,\ v,\ v_u)$. Let an optimal solution of (Level-1: *LPP*) be obtained as $(x^*,\ v_l^o,\ v^*,\ v_u^o)$.
Level 2: Construct the following scalar linear programming problem, namely Level-2: *LPP*,

$$\begin{aligned} \max \quad & (v_l,\ v_u) \\ \text{subject to,} \quad & \\ & v_l \leqq \sum_{i=1}^{m} (a_{ij})_l x_i^*, \ (j = 1,2,\ldots,n), \\ & v_u \leqq \sum_{i=1}^{m} (a_{ij})_u x_i^*, \ (j = 1,2,\ldots,n). \end{aligned}$$

Here v_l and v_u are decision variables. Since constraints for v_l and v_u are independent, the (Level-2: *LPP*) can be decomposed into the following linear programming problem

(Level-2: *LPP*1)

$$\begin{aligned} \max \quad & v_l \\ \text{subject to,} \quad & \\ & v_l \leqq \sum_{i=1}^{m} (a_{ij})_l x_i^*, \ (j = 1,2,\ldots,n) \end{aligned}$$

and,

(Level-2: *LPP*2)

$$\begin{aligned} \max \quad & v_u \\ \text{subject to,} \quad & \end{aligned}$$

$$v_u \leqq \sum_{i=1}^{m} (a_{ij})_u x_i^*, \ (j = 1, 2, \ldots, n).$$

Let optimal solutions of these two *LPP*'s be given by v_l^* and v_u^* respectively.

Level 3: Stop, as a Pareto optimal solution of (*MOP*1), namely $(x_1^*, x_2^*, v_l^*, v^*, v_u^*)$, has been obtained.

Remark 8.4.2. Once a Pareto optimal solution $\left(x^*, (v_l^*, v^*, v_u^*)\right)$ of (*MOP*1) has been obtained, the fuzzy game $FG = (S^m, S^n, \tilde{A})$ has been solved with x^* as an optimal strategy for Player I and $\tilde{v}^* = (v_l^*, v^*, v_u^*)$ as (fuzzy) value for Player I. Similar arguments hold for Player II as well and a two level linear programming approach for solving (*MOP*2) can be described to get an optimal strategy y^*, and a fuzzy value $\tilde{w}^* = (w_l^*, w^*, w_u^*)$ for Player II.

We now illustrate the above procedure with the help of following numerical example.

Example 8.4.3. (Li and Yang [42], Campos [10]). Consider the fuzzy matrix game $FG = (S^2, S^2, \tilde{A})$, where

$$\tilde{A} = \begin{pmatrix} (175, 180, 190) & (150, 156, 158) \\ (80, 90, 100) & (175, 180, 190) \end{pmatrix}.$$

As per our discussion above, to solve the fuzzy matrix game *FG*, we have to solve following multiobjective linear programming problems (*MOP*1) and (*MOP*2) in the sense of Pareto optimality

(*MOP*1) max (v_l, v, v_u)

subject to,

$$\begin{aligned} 175x_1 + 80x_2 &\geqq v_l \\ 150x_1 + 175x_2 &\geqq v_l \\ 180x_1 + 90x_2 &\geqq v \\ 156x_1 + 180x_2 &\geqq v \\ 190x_1 + 100x_2 &\geqq v_u \\ 158x_1 + 190x_2 &\geqq v_u \\ x_1 + x_2 &= 1 \\ x_1, x_2 &\geq 0, \end{aligned}$$

and

(*MOP*2) min (w_l, w, w_u)

subject to,

$$
\begin{aligned}
175y_1 + 150y_2 &\leqq w_l \\
80y_1 + 175y_2 &\leqq w_l \\
180y_1 + 156y_2 &\leqq w \\
90y_1 + 180y_2 &\leqq w \\
190y_1 + 158y_2 &\leqq w_u \\
100y_1 + 190y_2 &\leqq w_u \\
y_1 + y_2 &= 1 \\
y_1, y_2 &\geq 0.
\end{aligned}
$$

We now solve ($MOP1$) by the two level linear programming approach. For this, we first consider the (Level-1: LPP) as follows

$$
\begin{aligned}
&\max \quad v \\
&\text{subject to,} \\
&175x_1 + 80x_2 \geqq v_l \\
&150x_1 + 175x_2 \geqq v_l \\
&180x_1 + 90x_2 \geqq v \\
&156x_1 + 180x_2 \geqq v \\
&190x_1 + 100x_2 \geqq v_u \\
&158x_1 + 190x_2 \geqq v_u \\
&x_1 + x_2 = 1 \\
&x_1, x_2 \geq 0,
\end{aligned}
$$

This (Level-1: LPP) can be solved by the simplex algorithm to obtain its optimal solution $x^* = (0.7895,\ 0.2105)$, $v^* = 161.05$, $v_l^o = 61.398$ and $v_u^o = 163.63$.

Next we construct two Level-2 linear programming problems as follows

(Level-2: $LPP1$)

$$
\begin{aligned}
&\max \quad v_l \\
&\text{subject to,} \\
&v_l \leqq 175(0.7895) + 80(0.2105), \\
&v_l \leqq 150(0.7895) + 175(0.2105),
\end{aligned}
$$

and

(Level-2: $LPP2$)

$$
\begin{aligned}
&\max \quad v_u \\
&\text{subject to,} \\
&v_u \leqq 190(0.7895) + 158(0.2105), \\
&v_u \leqq 158(0.7895) + 190(0.2105).
\end{aligned}
$$

These Level-2 LPP's can further be simplified to following two linear programming problems

$$\begin{aligned} &\max \quad v_l \\ &\text{subject to,} \\ &\qquad v_l \leqq 155.0025, \\ &\qquad v_l \leqq 155.2625, \end{aligned}$$

and,

$$\begin{aligned} &\max \quad v_u \\ &\text{subject to,} \\ &\qquad v_u \leqq 183.264, \\ &\qquad v_u \leqq 164.736, \end{aligned}$$

whose optimal solutions are $v_l^* = 155.0025$ and $v_u^* = 164.736$ respectively. Therefore (x^*, v^*) with $x^* = (0.7895, 0.2105)$ and $\tilde{v}^* = (v_l^*, v^*, v_u^*) = (155.0025, 161.05, 164.736)$ is a Parteo optimal solution of $(MOP1)$.
Similarly, (y^*, w^*) with $y^* = (0.2105, 0.7895)$ and $\tilde{w}^* = (w_l^*, w^*, w_u^*) = (155.264, 161.05, 171.052)$ is a Pareto optimal solution of $(MOP2)$.

From the above discussion we conclude that the given fuzzy game has optimal strategies for Player I and Player II as $(0.7895, 0.2105)$ and $(0.2105, 0.7895)$ respectively, and, values of the game for Player I and Player II are $(155.0025, 161.05, 164.736)$ and $(155.264, 161.05, 171.052)$ respectively.

Here it must be noted that because the given matrix game is fuzzy, Player I and Player II will have fuzzy values only as indicated above. Further, we also know the complete membership function of $\tilde{v}^*$ and $\tilde{w}^*$, unlike the earlier results obtained in Chapter 7 where only representative values $F(\tilde{v}^*)$ and $F(\tilde{w}^*)$ were obtained.

8.5 Conclusions

Continuing with our discussion from Chapter 7, in this chapter we have presented Maeda's model and Li's model along with Li and Yang's model for matrix games with fuzzy pay-offs. These model do not require any priori choice of the ranking function and therefore seemingly are more useful for the fuzzy scenario. However, unlike the ranking function approach, these models do not seem to provide any equivalence in terms of duality in fuzzy linear programming.

9

Fuzzy Bi-Matrix Games

9.1 Introduction

In this chapter, bi-matrix games in fuzzy scenario are considered and an attempt is made to conceptualize the meaning of an *equilibrium solution* for such games. The fuzzy scenario could be in terms of fuzzy goals or fuzzy pay-offs, or both. It is assumed that each player tries to maximize some "measure" of attainment for his goal and the aim is to find an equilibrium solution with respect to that "measure" of attainment of these fuzzy goals. As in fuzzy linear programming and fuzzy matrix games, this analysis will lead to an equivalent (crisp) mathematical programming problem whose solution will render the desired equilibrium solution for the fuzzy case.

The contents of this chapter are based on Maeda [49], Nishizaki and Sakawa [61] and Vijay, Chandra and Bector [77]. This chapter is divided into five main sections, namely *bi-matrix games with fuzzy goals: Nishizaki and Sakawa's model, bi-matrix games with fuzzy goals: another approach, bi-matrix games with fuzzy pay-offs: a ranking function approach, bi-matrix games with fuzzy goals and fuzzy pay-offs,* and *bi-matrix games with fuzzy pay-offs: a possibility measure approach.*

9.2 Bi-matrix games with fuzzy goals: Nishizaki and Sakawa's model

Nishizaki and Sakawa [61] defined a bi-matrix game with fuzzy goals as

$$BFG = (S^m,\ S^n,\ A,\ B,\ \mu_{G_1},\ \mu_{G_2},\ \underline{a},\ \bar{a},\ \underline{b},\ \bar{b}),$$

where S^m, S^n, A and B are as defined in Chapter 1, and $\underline{a}$, $\bar{a}$ ($\bar{a} > \underline{a}$) are scalars which are used in defining the fuzzy goal G_1 and it's associated membership function μ_{G_1} for Player I. In a similar manner, the scalars $\underline{b}$, $\bar{b}$ ($\bar{b} > \underline{b}$) and the membership function μ_{G_2} are to be understood for Player II.

Although $\underline{a}$, $\bar{a}$, $\underline{b}$ and $\bar{b}$ could be any scalars with $\bar{a} > \underline{a}$ and $\bar{b} > \underline{b}$; Nishizaki and Sakawa [61] has suggested $\underline{a} = \min_i \min_j a_{ij}$, $\bar{a} = \max_i \max_j a_{ij}$, $\underline{b} = \min_i \min_j b_{ij}$ and $\bar{b} = \max_i \max_j b_{ij}$ with the interpretation that $\underline{a}$ (respectively $\underline{b}$) gives the worst degree of satisfaction and $\bar{a}$ (respectively $\bar{b}$) gives the best degree of satisfaction for Player I (respectively Player II).

Definition 9.2.1 (Fuzzy goal for Player I). *Let $D_1 = \{x^T Ay : x \in S^m,\ y \in S^n\}$ and $v \in D_1$. Then a fuzzy goal for Player I is a fuzzy set G_1 with the membership function $\mu_{G_1} : D_1 \longrightarrow [0,1]$. The fuzzy goal G_1 can also be symbolically written as $v \gtrsim_{p_0} \bar{a}$, where the tolerance p_0 equals $(\bar{a} - \underline{a})$.*

Definition 9.2.2 (Fuzzy goal for Player II). *Let $D_2 = \{x^T By : x \in S^m,\ y \in S^n\}$ and $w \in D_2$. Then a fuzzy goal for Player II is a fuzzy set G_2 with the membership function $\mu_{G_2} : D_2 \longrightarrow [0,1]$. The fuzzy goal G_2 can also be symbolically written as $w \gtrsim_{q_0} \bar{b}$, where the tolerance q_0 equals $(\bar{b} - \underline{b})$.*

Now on wards for the sake of convenience, we shall write μ_1 and μ_2 for μ_{G_1} and μ_{G_2} respectively.

Definition 9.2.3 (Equilibrium solution of *BFG*). *A pair $(x^*, y^*) \in S^m \times S^n$ is said to be an equilibrium solution of BFG if*

$$\mu_1(x^{*T}Ay^*) \geq \mu_1(x^T Ay^*),\ \forall\, x \in S^m,$$

and

$$\mu_2(x^{*T}By^*) \geq \mu_2(x^{*T}By),\ \forall\, y \in S^n.$$

Thus an equilibrium solution for *BFG* is defined with respect to the degree of attainment of the fuzzy goals.

If the membership functions of the fuzzy goals G_1 and G_2 are linear then they can be represented as

$$\mu_1(x^T Ay) = \begin{cases} 0 & , x^T Ay \leq \underline{a}, \\ 1 - \dfrac{\bar{a} - x^T Ay}{\bar{a} - \underline{a}} & , \underline{a} < x^T Ay < \bar{a}, \\ 1 & , x^T Ay \geq \bar{a}, \end{cases}$$

and

$$\mu_2(x^T By) = \begin{cases} 0 & , x^T By \leq \underline{b}, \\ 1 - \dfrac{\bar{b} - x^T By}{\bar{b} - \underline{b}} & , \underline{b} < x^T By < \bar{b}, \\ 1 & , x^T By \geq \bar{b}. \end{cases}$$

Now,

$$1 - \left(\frac{\bar{a} - x^T Ay}{\bar{a} - \underline{a}} \right) = \left(\frac{-\underline{a}}{\bar{a} - \underline{a}} \right) + x^T \left(\frac{A}{\bar{a} - \underline{a}} \right) y$$

and

$$1 - \left(\frac{\bar{b} - x^T By}{\bar{b} - \underline{b}} \right) = \left(\frac{-\underline{b}}{\bar{b} - \underline{b}} \right) + x^T \left(\frac{B}{\bar{b} - \underline{b}} \right) y.$$

Therefore letting

$$\hat{A} = \frac{A}{\bar{a} - \underline{a}}, \ \hat{B} = \frac{B}{\bar{b} - \underline{b}}, \ c_1 = -\frac{\underline{a}}{\bar{a} - \underline{a}} \text{ and } c_2 = -\frac{\underline{b}}{\bar{b} - \underline{b}},$$

the membership functions $\mu_1(x^T Ay)$ and $\mu_2(x^T By)$ can be rewritten as

$$\mu_1(x^T Ay) = \begin{cases} 0 & , x^T Ay \leq \underline{a}, \\ c_1 + x^T \hat{A} y & , \underline{a} \leq x^T Ay \leq \bar{a}, \\ 1 & , \bar{a} \leq x^T Ay, \end{cases}$$

and

$$\mu_2(x^T By) = \begin{cases} 0 & , x^T By \leq \underline{b}, \\ c_2 + x^T \hat{B} y & , \underline{b} \leq x^T By \leq \bar{b}, \\ 1 & , \bar{b} \leq x^T By, \end{cases}$$

respectively.

We now consider the (crisp) bi-matrix games $BG = (S^m, S^n, A, B)$ and $\widehat{BG} = (S^m, S^n, \hat{A}, \hat{B})$. The following theorem states that bi-matrix games BG and $\widehat{BG}$ have the same set of equilibrium solutions.

Theorem 9.2.1 *The pair* $(x^*, y^*) \in S^m \times S^n$ *is an equilibrium solution of* $BG = (S^m, S^n, A, B)$ *if and only if it is also an equilibrium solution of* $\widehat{BG} = (S^m, S^n, \hat{A}, \hat{B})$.

Proof. First, we will prove that a pair of strategies (x^*, y^*) which is an equilibrium solution of BG is also an equilibrium solution of $\widehat{BG}$. Now (x^*, y^*) being an equilibrium solution of BG, we have

$$x^{*T}Ay^* \geq x^T Ay^*, \ \forall\, x \in S^m,$$

and

$$x^{*T}By^* \geq x^{*T}By, \ \forall\, y \in S^n.$$

This implies

$$x^{*T}\left(\frac{A}{\bar{a}-\underline{a}}\right)y^* \geq x^T\left(\frac{A}{\bar{a}-\underline{a}}\right)y^*, \ \forall\, x \in S^m,$$

and

$$x^{*T}\left(\frac{B}{\bar{b}-\underline{b}}\right)y^* \geq x^{*T}\left(\frac{B}{\bar{b}-\underline{b}}\right)y, \ \forall\, y \in S^n,$$

which gives that (x^*, y^*) is an equilibrium solution of $\widehat{BG}$. The converse follows by just going backward.

Our next theorem states that an equilibrium solution of the (crisp) bi-matrix game BG (or $\widehat{BG}$) is also an equilibrium solution with respect to the degree of attainment of the fuzzy goals for the fuzzy bi-matrix game BFG.

Theorem 9.2.2 *Let (x^*, y^*) be an equilibrium solution of the (crisp) bi-matrix game BG (or $\widehat{BG}$). Also let the membership functions μ_1 and μ_2 for fuzzy goals G_1 and G_2 be linear as described above. Then (x^*, y^*) is also an equilibrium solution of the fuzzy bi-matrix game BFG in the sense of Definition 9.2.3.*

Proof. Let (x^*, y^*) be an equilibrium solution of the (crisp) bi-matrix game BG. Then by Theorem 9.2.1 it is also an equilibrium solution of the (crisp) bi-matrix game $\widehat{BG}$. Therefore we have

$$x^{*T}Ay^* \geq x^T Ay^*, \ \forall\, x \in S^m,$$

$$x^{*T}By^* \geq x^{*T}By, \ \forall\, y \in S^n,$$

$$x^{*T}\hat{A}y^* \geq x^T \hat{A}y^*, \ \forall\, x \in S^m,$$

and

$$x^{*T}\hat{B}y^* \geq x^{*T}\hat{B}y, \ \forall\, y \in S^n.$$

We also recall that

$$\mu_1(x^T A y) = \begin{cases} 0 & , x^T A y \le \underline{a}, \\ c_1 + x^T \hat{A} y & , \underline{a} \le x^T A y \le \bar{a}, \\ 1 & , \bar{a} \le x^T A y. \end{cases}$$

Now we consider the following cases and observe that

Case 1. If $x^T A y^* \le x^{*T} A y^* \le \underline{a}$, then

$$\mu_1(x^{*T} A y^*) = \mu_1(x^T A y^*) = 0.$$

Case 2. If $x^T A y^* \le \underline{a} \le x^{*T} A y^*$, then

$$\mu_1(x^{*T} A y^*) \ge \mu_1(x^T A y^*) = 0.$$

Case 3. If $\underline{a} \le x^T A y^* \le x^{*T} A y^* \le \bar{a}$, then

$$\mu_1(x^T A y^*) = x^T \hat{A} y^* + c_1,$$
$$\mu_1(x^{*T} A y^*) = x^{*T} \hat{A} y^* + c_1,$$

which because of $x^{*T} \hat{A} y^* \ge x^T \hat{A} y^*$, gives

$$\mu_1(x^{*T} A y^*) \ge \mu_1(x^T A y^*).$$

Case 4. If $x^T A y^* \le \bar{a} \le x^{*T} A y^*$, then

$$\mu_1(x^T A y^*) \le 1 \text{ and } \mu_1(x^{*T} A y^*) = 1.$$

Therefore

$$1 = \mu_1(x^{*T} A y^*) \ge \mu_1(x^T A y^*).$$

Case 5. If $\bar{a} \le x^T A y^* \le x^{*T} A y^*$, then

$$\mu_1(x^{*T} A y^*) = \mu_1(x^T A y^*) = 1.$$

The above cases imply that for all $x \in S^m$, $\mu_1(x^{*T} A y^*) \ge \mu_1(x^T A y^*)$. Similarly for all $y \in S^n$, we have $\mu_2(x^{*T} B y^*) \ge \mu_2(x^T B y^*)$.

Now similar to the crisp case, we have the following theorem which states that if (x^*, y^*) is an optimal solution of the quadratic programming problem $\widehat{QPP}$ then (x^*, y^*) is an equilibrium solution of the fuzzy bi-matrix game BFG.

Theorem 9.2.3 *Let (x^*, y^*, p^*, q^*) be an optimal solution of the following quadratic programming problem $\widehat{BFG}$*

$$\begin{aligned} \max \quad & x^T \hat{A} y + x^T \hat{B} y - p - q \\ \text{subject to,} \quad & \\ & \hat{A} y \le pe, \\ & \hat{B}^T x \le qe, \\ & e^T x = 1, \\ & e^T y = 1, \\ & x \ge 0, \\ & y \ge 0. \end{aligned}$$

Then (x^*, y^*) is an equilibrium solution of the fuzzy bi-matrix game (BFG).

Proof. Given that (x^*, y^*, p^*, q^*) is a solution of $\widehat{QPP}$, Theorem 1.6.1 implies that (x^*, y^*) is an equilibrium solution of the (crisp) bi-matrix game $\widehat{BG} = (S^m, S^n, \hat{A}, \hat{B})$. But, by Theorem 9.2.2, this means that (x^*, y^*) is an equilibrium solution of the fuzzy bi-matrix game *BFG* in the sense of Definition 9.2.3.

Remark 9.2.1. Once an optimal solution (x^*, y^*, p^*, q^*) of $\widehat{QPP}$ has been obtained, (x^*, y^*) gives an equilibrium solution of the fuzzy bi-matrix game (*BFG*). The degree of attainment of fuzzy goals G_1 and G_2 can then be determined by evaluating $x^{*T}Ay^*$ (or $x^{*T}\hat{A}y^*$) and $x^{*T}By^*$ (or $x^{*T}\hat{B}y^*$) and then employing the membership functions μ_1 and μ_2.

9.3 Bi-matrix games with fuzzy goals: another approach

In the last section, we have presented Nishizaki and Sakawa's [61] approach to study bi-matrix games with fuzzy goals *BFG*. In this section, we conceptualize such a game in a manner which is somewhat different than that of Nishizaki and Sakawa [61] and show its equivalence to a special type of nonlinear programming problem in which the objective function as well as all constraint functions are linear except two quadratic constraint functions. This new bi-matrix game with fuzzy goals is denoted by *BGFG* and the results reported here for *BGFG* are based on Vijay, Chandra and Bector [77].

Let S^m, S^n, A and B be as introduced in the Section 9.2. Let v_0, w_0 be scalars representing the aspiration levels of Player I and Player II respectively. Then the bi-matrix game with fuzzy goals under consideration here is, denoted by *BGFG*, and is defined as

$$BGFG = (S^m, S^n, A, B, v_0, \gtrsim, w_0, \lesssim),$$

where '$\lesssim$' and '$\gtrsim$' are the fuzzified versions of '$\leq$' and '$\geq$' respectively. Therefore the game *BGFG* gets fixed only when the specific choices of membership functions are made to define fuzzy inequalities '$\gtrsim$' and '$\lesssim$'. Here we shall interpret '$\gtrsim$' and '$\lesssim$' in the sense of Zimmerman [91] although a more general interpretation in terms of modalities and fuzzy relations can also be taken.

Let t be a real variable and $a \in \mathbb{R}$. Let $\hat{p} > 0$. We now recall the notation "$t \gtrsim_{\hat{p}} a$" from Chapter 6 and note that this is to be read as "t is essentially greater or equal to a with tolerance $\hat{p}$", and is to be understood in terms of the following membership function

$$\mu_D(t) = \begin{cases} 1 & , t \geq a, \\ 1 - \left(\dfrac{a-t}{\hat{p}}\right) & , (a - \hat{p}) \leq t \leq a, \\ 0 & , t < (a - \hat{p}). \end{cases}$$

We also recall the below given lemma from Chapter 6.

Lemma 9.3.1. *Let* $t_1 \gtrsim_{\hat{p}} a$, $t_2 \gtrsim_{\hat{p}} a$, $\alpha \geq 0$, $\beta \geq 0$ *and* $\alpha + \beta = 1$. *Then* $\alpha t_1 + \beta t_2 \gtrsim_{\hat{p}} a$.

In view of the above discussion we include tolerances p_0, p_0' for Player I, and, q_0 and q_0' for Player II respectively in our definition of the fuzzy bi-matrix game $BGFG$ and therefore take $BGFG$ as

$$BGFG = (S^m,\ S^n,\ A,\ B,\ v_0,\ p_0,\ p_0',\ w_0,\ q_0,\ q_0',\ \lesssim,\ \gtrsim).$$

Now we define the meaning of an "equilibrium solution" of the fuzzy bi-matrix game $BGFG$.

Definition 9.3.1 (Equilibrium solution of $BGFG$**).** *A point* $(x^*,\ y^*) \in S^m \times S^n$ *is called an equilibrium solution of the fuzzy bi-matrix game* $BGFG$ *if*

$$\begin{aligned} x^T A y^* &\lesssim_{p_0} v_0,\ \forall\ x \in S^m, \\ x^{*T} B y &\lesssim_{q_0} w_0,\ \forall\ y \in S^n, \\ x^{*T} A y^* &\gtrsim_{p_0'} v_0, \\ x^{*T} B y^* &\gtrsim_{q_0'} w_0. \end{aligned}$$

Remark 9.3.2. For the crisp scenario above inequalities become

$$x^T A y^* \leq v_0, \forall x \in S^m,$$

$$x^{*T} B y \leq w_0, \forall y \in S^n,$$

$$x^{*T} A y^* \geq v_0,$$

and

$$x^{*T} B y^* \geq w_0.$$

Therefore for all $x \in S^m$, $y \in S^n$ we have $x^T A y^* \leq x^{*T} A y^*$ and $x^{*T} B y \leq x^{*T} B y^*$ which is the same as the definition for an equilibrium solution for the (crisp) bi-matrix game BG.

In the Definition 9.3.1, the sets S^m and S^n are convex polytopes, therefore in view of Lemma 9.3.1, for the specific choice of membership functions of type $\mu_D(t)$, it is sufficient to consider only the extreme points (i.e. pure strategies) of S^m and S^n. This observation leads to the following fuzzy non-linear programming problem (*FNP*)

(*FNP*) Find (x, y) such that

$$
\begin{aligned}
A_i y &\lesssim_{p_0} v_0, \ (i = 1, 2, \ldots, m), \\
{B_j}^T x &\lesssim_{q_0} w_0, \ (j = 1, 2, \ldots, n), \\
x^T A y &\gtrsim_{p'_0} v_0, \\
x^T B y &\gtrsim_{q'_0} w_0, \\
x &\in S^m, \\
y &\in S^n,
\end{aligned}
$$

where for $i = 1, 2, \ldots, m$, A_i denotes the ith row of A and for $j = 1, 2, , \ldots, n$, B_j denotes the jth column of B.

Now as per the requirement for the use of Lemma 9.3.1, we have to define the specific linear membership functions of type $\mu_D(t)$ for all the fuzzy constraints. Therefore membership function $\mu_i(A_i y)$, $(i = 1, 2, \ldots, m)$, which gives the degree to which y satisfies fuzzy constraint $A_i y \lesssim_{p_o} v_o$ and $\nu_j(B_j^T x)$, $(j = 1, 2, \ldots, n)$, which gives the degree to which x satisfies the fuzzy constraint $B_j^T x \lesssim_{p_o} w_o$, are given as

$$
\mu_i(A_i y) = \begin{cases} 1 & , A_i y \leq v_0, \\ 1 - \dfrac{A_i y - v_0}{p_0} & , v_0 \leq A_i y \leq v_0 + p_0, \\ 0, & , A_i y \geq v_0 + p_0, \end{cases}
$$

and

$$
\nu_j({B_j}^T x) = \begin{cases} 1 & , {B_j}^T x \leq w_0, \\ 1 - \dfrac{{B_j}^T x - w_0}{q_0} & , w_0 \leq {B_j}^T x \leq w_0 + q_0, \\ 0 & , {B_j}^T x \geq w_0 + q_0, \end{cases}
$$

respectively.

Similarly, linear membership functions for the fuzzy constraints $x^T A y \gtrsim_{p'_0} v_0$ and $x^T B y \gtrsim_{q'_0} w_0$ are defined as follows

$$
\mu_0(x^T A y) = \begin{cases} 1 & , x^T A y \geq v_0, \\ 1 - \dfrac{v_0 - x^T A y}{p'_0} & , v_0 \geq x^T A y \geq v_0 - p'_0, \\ 0 & , x^T A y \geq v_0 - p'_0, \end{cases}
$$

and

$$\nu_0(x^T By) = \begin{cases} 1 & , x^T By \geq w_0, \\ 1 - \dfrac{w_0 - x^T By}{q_0'} & , w_0 \geq x^T By \geq w_0 - q_0', \\ 0, & , x^T By \leq w_0 - q_0'. \end{cases}$$

Now employing the above mentioned membership functions and following Zimermann's approach [90], we obtain the crisp equivalent of the fuzzy non-linear programming (*FNP*) as

(*NLP*) max λ

subject to,

$$\lambda \leq 1 - \frac{A_i y - v_0}{p_0}, \quad (i = 1, 2, \dots, m),$$

$$\lambda \leq 1 - \frac{B_j^T x - w_0}{q_0}, \quad (j = 1, 2, \dots, n),$$

$$\lambda \leq 1 + \frac{x^T Ay - v_0}{p_0'},$$

$$\lambda \leq 1 + \frac{x^T By - w_0}{q_0'},$$

$$x \in S^m,$$

$$y \in S^n,$$

$$\lambda \in [0, 1].$$

The above discussion leads to the following theorem:

Theorem 9.3.1 *Let (x^*, y^*, λ^*) be an optimal solution to the problem (NLP). Then (x^*, y^*) is an equilibrium solution of the fuzzy bi-matrix game BGFG and λ^* is the least degree up to which the respective aspiration levels (goals) v_0 and w_0 of Player I and Player II are met.*

Remark 9.3.3. Let (x^*, y^*, λ^*) be a solution to the (*NLP*) with $\lambda^* = 1$. Then *BGFG* becomes the (crisp) bi-matrix game *BG* and (x^*, y^*) becomes its equilibrium solution. Therefore various results of (crisp) bi-matrix game theory follow as a special case of fuzzy bi-matrix game theory. Further, for $\lambda^* = 1$, the non-linear programming problem (*NLP*) reduces to the system

$$
\begin{aligned}
A_i y &\le v_0, \ (i = 1, 2, \ldots, m),\\
{B_j}^T x &\le w_0, (j = 1, 2, \ldots, n),\\
x^T A y &= v_0,\\
x^T B y &= w_0,\\
x &\in S^m,\\
y &\in S^n,
\end{aligned}
$$

which implies Theorem 1.6.1.

Special Cases:

It has already been explained in Remark 9.3.3 that various results of crisp bi-matrix game theory follow as a special case of fuzzy bi-matrix game theory. In the following certain other special cases are presented so as to bring out differences/similarities between the results of this section with that of Nishizaki and Sakawa as presented in Section 9.2.

1. Let us recall from Section 9.2 that if (x^*, y^*, p^*, q^*) is a solution of the quadratic programming problem (QPP) then it is also an equilibrium solution of the fuzzy bi-matrix game BFG, where

   (QPP): max $\quad x^T\left(\dfrac{A}{\bar{a} - \underline{a}}\right)y + x^T\left(\dfrac{B}{\bar{b} - \underline{b}}\right)y - p - q$

   subject to,

$$
\begin{aligned}
\left(\frac{A}{\bar{a} - \underline{a}}\right)y &\le pe,\\
\left(\frac{B^T}{\bar{b} - \underline{b}}\right)x &\le qe,\\
e^T x &= 1,\\
e^T y &= 1,\\
x, y &\ge 0.
\end{aligned}
$$

   Now from Mangasarian and Stone [52] and also from results of Section 9.2, it is known that if (x^*, y^*, p^*, q^*) is an optimal solution to the above quadratic programming problem (QPP) then ${x^*}^T A y^* = p(\bar{a} - \underline{a})$ and ${x^*}^T B y^* = q(\bar{b} - \underline{b})$. Therefore to obtain (crisp) bi-matrix game as a special case of the fuzzy bi-matrix game (BFG), the membership function values $\mu_1({x^*}^T A y^*)$ and $\mu_2({x^*}^T B y^*)$ should be equal to 1. Therefore we should have ${x^*}^T A y^* \ge \bar{a}$ and ${x^*}^T B y^* \ge \bar{b}$; which can be written as $p(\bar{a} - \underline{a}) \ge \bar{a}$ and $q(\bar{b} - \underline{b}) \ge \bar{b}$. Since no relationship between $\bar{a}$ and $p(\bar{a} - \underline{a})$, and also between $\bar{b}$ and $q(\bar{b} - \underline{b})$, is given in Nishizaki and Sakawa [61] the results of crisp bi-matrix game do not seem to follow from the fuzzy bi-matrix game *BFG*.

2. As has been observed in Chapter 1, a crisp two person zero-sum matrix game is a special case of crisp bi-matrix game. More explicitly, the quadratic programming problem given by Mangasarian and Stone [52] decomposes itself into a pair of two primal-dual linear programming problems for the case $A = -B$. Since assumption $A = -B$ does not imply $\frac{A}{\bar{a} - \underline{a}} = \frac{B}{\bar{b} - \underline{b}}$ in general the above quadratic programming problem (QPP) given by Nishizaki and Sakawa may not always decompose in this manner. There is similar difficulty with the fuzzy bi-matrix $(BGFG)$ as well. However if we choose $\bar{a} = \max_i \max_j a_{ij}$, $\underline{a} = \min_i \min_j a_{ij}$, $\bar{b} = \max_i \max_j b_{ij}$ and $\underline{b} = \min_i \min_j b_{ij}$ then (QPP) decomposes itself into two linear programming problems which are dual to each other. In a similar manner if $\lambda^* = 1$, $\alpha_o = -\beta_o$ then for the case $B = -A$, an optimal solution $(x^*, y^*, \lambda^* = 1)$ gives $x^T A y^* \leq x^{*^T} A y^* \leq x^{*^T} A y$ for all $x \in S^m$ and $y \in S^n$, thereby implying that (x^*, y^*) is a saddle point of the (crisp) matrix game $G = (S^m, S^n, A)$.

9.4 Bi-matrix games with fuzzy pay-offs: a ranking function approach

In Chapter 7, we have already seen a ranking function approach to matrix games with fuzzy pay-offs. Here we attempt to extend the same to bi-matrix games with fuzzy pay-offs . Let S^m, S^n be as in Section 9.2 and, $\tilde{A}$ and $\tilde{B}$ be the pay-off matrices with entries as fuzzy numbers for Player I and Player II respectively. Then *a bi-matrix game with fuzzy pay-offs*, denoted by *BGFP*, is defined as

$$BGFP = (S^m, S^n, \tilde{A}, \tilde{B}).$$

Now, we define the meaning of an "equilibrium solution" of the fuzzy bi-matrix game *BGFP*. For this we first have the following definition.

Definition 9.4.1 (Reasonable solution of ***BGFP*****).** *Let $\tilde{v}$, $\tilde{w} \in N(\mathbb{R})$. Then $(\tilde{v}, \tilde{w})$ is called a reasonable solution of the fuzzy bi-matrix game BGFP if there exists $x^* \in S^m$, $y^* \in S^n$ such that*

$$x^T \tilde{A} y^* \lesssim_{\tilde{p}} \tilde{v},\ \forall x \in S^m,$$
$$x^{*T} \tilde{B} y \lesssim_{\tilde{q}} \tilde{w},\ \forall y \in S^n,$$
$$x^{*T} \tilde{A} y^* \gtrsim_{\tilde{p}'} \tilde{v},$$
$$x^{*T} \tilde{B} y^* \gtrsim_{\tilde{q}'} \tilde{w}.$$

If ($\tilde{v}$, $\tilde{w}$) is a reasonable solution of BGFP then $\tilde{v}$ (respectively $\tilde{w}$) is called a reasonable value for Player I (respectively Player II).

Definition 9.4.2 (Equilibrium solution of *BGFP*). *Let T_1 and T_2 be the set of all reasonable values $\tilde{v}$ and $\tilde{w}$ for Player I and Player II respectively where $\tilde{v}$, $\tilde{w} \in N(\mathbb{R})$. Let there exist $\tilde{v}^* \in T_1$, $\tilde{w}^* \in T_2$ such that*

$$F(\tilde{v}^*) \geq F(\tilde{v}),\ \forall \tilde{v} \in T_1,$$

and

$$F(\tilde{w}^*) \geq F(\tilde{w}), \forall \tilde{w} \in T_2,$$

where $F : N(\mathbb{R}) \to \mathbb{R}$ is the chosen defuzzification function. Then the pair (x^*, y^*) is called an equilibrium solution of the game BGFP. Also $\tilde{v}^*$ (respectively $\tilde{w}^*$) is called the value of the game BGFP for Player I (respectively Player II).

By using the above definitions for the game *BGFP*, we now construct the following fuzzy non-linear programming problem

$(FNP1)$ max $F(\tilde{v}) + F(\tilde{w})$

subject to,

$$x^T \tilde{A} y \lesssim_{\tilde{p}} \tilde{v}, \quad \forall x \in S^m,$$
$$x^T \tilde{B} y \lesssim_{\tilde{q}} \tilde{w}, \quad \forall y \in S^n,$$
$$x^T \tilde{A} y \gtrsim_{\tilde{p}'} \tilde{v},$$
$$x^T \tilde{B} y \gtrsim_{\tilde{q}'} \tilde{w},$$
$$x \in S^m,$$
$$y \in S^n,$$
$$\tilde{v}, \tilde{w} \in N(\mathbb{R}).$$

Now recalling the explanation of the double fuzzy constraints as explained in Chapter 7 and noting that the relations $\textcircled{\scriptsize$\geq$}$ and $\textcircled{\scriptsize$\leq$}$ preserve the ranking when fuzzy numbers are multiplied by positive scalars, it makes sense to consider only the extreme points of sets S^m and S^n in the constraints of $(FNP1)$. Therefore the above problem $(FNP1)$ will be converted into

$(FNP2)$ max $F(\tilde{v}) + F(\tilde{w})$

subject to,

$$\begin{aligned}
\tilde{A}_i y &\lesssim_{\tilde{p}} \tilde{v}, \quad (i = 1,2,\dots,m),\\
x^T \tilde{B}_j &\lesssim_{\tilde{q}} \tilde{w}, \quad (j = 1,2,\dots,n),\\
x^T \tilde{A} y &\gtrsim_{\tilde{p}'} \tilde{v},\\
x^T \tilde{B} y &\gtrsim_{\tilde{q}'} \tilde{w},\\
x &\in S^m,\\
y &\in S^n,\\
\tilde{v}, \tilde{w} &\in N(\mathbb{R}).
\end{aligned}$$

Here $\tilde{A}_i$ denotes the ith row of $\tilde{A}$ and $\tilde{B}_j$ denotes the jth column of $\tilde{B}$, $(i = 1,2,\dots,m;\ j = 1,2,\dots,n)$.

By using the resolution procedure for the double fuzzy constraints in $(FNP2)$ as discussed in Chapter 7, we obtain

$(FNP3)$ max $F(\tilde{v}) + F(\tilde{w})$

subject to,

$$\begin{aligned}
\sum_{j=1}^{n} \tilde{a}_{ij} y_j &\circledleq \tilde{v} + (1-\lambda)\tilde{p}, \quad (i = 1,2,\dots,m),\\
\sum_{i=1}^{m} \tilde{b}_{ij} x_i &\circledleq \tilde{w} + (1-\eta)\tilde{q}, \quad (j = 1,2,\dots,n),\\
x^T \tilde{A} y &\circledgeq \tilde{v} - (1-\lambda)\tilde{p}',\\
x^T \tilde{B} y &\circledgeq \tilde{w} - (1-\eta)\tilde{q}',\\
x &\in S^m,\\
y &\in S^n,\\
\lambda, \eta &\in [0,1],\\
\tilde{v}, \tilde{w} &\in N(\mathbb{R}).
\end{aligned}$$

Now by utilizing the chosen defuzzification function for the constraints in $(FNP3)$, the problem can further be written as

$(NLP1)$ max $F(\tilde{v}) + F(\tilde{w})$

subject to,

$$\begin{aligned}
\sum_{j=1}^{n} F(\tilde{a}_{ij}) y_j &\le F(\tilde{v}) + (1-\lambda)F(\tilde{p}), \quad (i = 1,2,\dots,m),\\
\sum_{i=1}^{m} F(\tilde{b}_{ij}) x_i &\le F(\tilde{w}) + (1-\eta)F(\tilde{q}), \quad (j = 1,2,\dots,n),\\
F(x^T \tilde{A} y) &\ge F(\tilde{v}) - (1-\lambda)F(\tilde{p}'),
\end{aligned}$$

$$\begin{aligned} F(x^T\tilde{B}y) \geq\ & F(\tilde{w}) - (1-\eta)F(\tilde{q}'), \\ & x \in S^m, \\ & y \in S^n, \\ & \lambda, \eta \in [0,1], \\ & \tilde{v}, \tilde{w} \in N(\mathbb{R}). \end{aligned}$$

From the above discussion we observe that for solving the fuzzy bi-matrix game *BGFP* we have to solve the crisp non-linear programming problem (*NLP*1). Also, if $(x^*, \lambda^*, \tilde{v}^*, y^*, \eta^*, \tilde{w}^*)$ is an optimal solution of the crisp non-linear programming problem (*NLP*1), then (x^*, y^*) is an equilibrium solution of the game *BGFP*.

These results can now be summarized in the form of the following theorem.

Theorem 9.4.1 *The fuzzy bi-matrix game BGFP described by BGFP =* $(S^m, S^n, \tilde{A}, \tilde{B})$ *is equivalent to the crisp non-linear programming problem (NLP1) in which the objective as well as all constraint functions are linear except two constraint functions, which are quadratic.*

Remark 9.4.1. If all the fuzzy numbers are to be taken as crisp numbers i.e. $\tilde{a}_{ij} = a_{ij}$, $\tilde{b}_{ij} = b_{ij}$, $\tilde{v} = v$, $\tilde{w} = w$ and in the optimal solution of (*NLP*1), $\lambda^* = \eta^* = 1$, then the fuzzy game *BGFP* reduces to the crisp bi-matrix game *BG*. Thus if $\tilde{A}$, $\tilde{B}$, $\tilde{v}$ and $\tilde{w}$ are crisp and $\lambda^* = \eta^* = 1$, then *BGFP* reduces to *BG* and the crisp non-linear programming problem (*NLP*1) reduces to the non-linear programming problem (*NLP*2)

(*NLP*2)

$$\begin{aligned} \max \quad & v + w \\ \text{subject to,} \quad & \\ & Ay \leq ve, \\ & B^T x \leq we, \\ & x^T A y \geq v, \\ & x^T B y \geq w, \\ & x \in S^m, \\ & y \in S^n, \\ & v, w \in \mathbb{R}. \end{aligned}$$

For the case $B = -A$, the bi-matrix game *BG* reduces to the matrix game $G = (S^m, S^n, A)$ and the problem (*NLP*2) reduces to the system

$$
\begin{aligned}
Ay &\leq ve, \\
A^T x &\geq -we, \\
v + w &= 0, \\
x &\in S^m, \\
y &\in S^n, \\
v, w &\in \mathbb{R}.
\end{aligned}
$$

The above system is equivalent to the usual primal-dual pair of linear programming problems corresponding to the matrix game $G = (S^m, S^n, A)$.

Remark 9.4.2. In general it will be difficult to solve the problem ($NLP1$) and obtain exact membership functions for fuzzy values $\tilde{v}^*$ and $\tilde{w}^*$ because the decision variables $\tilde{v}$ and $\tilde{w}$ are fuzzy and their representation will involve large number of parameters. For example, if $\tilde{v}$ is a TFN (v_l, v, v_u) then to determine $\tilde{v}$ completely we need all of these three variables. Therefore, from the computational point of view it becomes necessary to take $F(\tilde{v})$ and $F(\tilde{w})$ as real variables V and W respectively and modify problem ($NLP1$) as follows

($NLP3$) max $V + W$

subject to,

$$
\begin{aligned}
&\sum_{j=1}^{n} F(\tilde{a}_{ij}) y_j \leq V + (1-\lambda) F(\tilde{p}), \quad (i = 1, 2, \ldots, m), \\
&\sum_{i=1}^{m} F(\tilde{b}_{ij}) x_i \leq W + (1-\eta) F(\tilde{q}), \quad (j = 1, 2, \ldots, n), \\
&F(x^T \tilde{A} y) \geq V - (1-\lambda) F(\tilde{p}'), \\
&F(x^T \tilde{B} y) \geq W - (1-\eta) F(\tilde{q}'), \\
&x \in S^m, \\
&y \in S^n, \\
&V, W \in \mathbb{R}. \\
&\lambda, \eta \in [0, 1].
\end{aligned}
$$

In this situation, in spite of knowing that the "values" for Player I and Player II are fuzzy with appropriate membership functions, we shall only get numerical values V^* and W^* for Player I and Player II respectively and the actual fuzzy value for Player I and Player II will be "close to" V^* and W^* respectively. Thus we shall not get exact membership functions for the fuzzy values of Player I and Player II even though these may be very much desirable. In the particular case when F

is Yager's first index [87], the numerical values V^* (respectively W^*) will represent the "centroid" or "average" value for Player I (respectively Player II).

9.5 Bi-matrix Game with Fuzzy Goals and Fuzzy Pay-offs

Let S^m, S^n, $\tilde{A}$ and $\tilde{B}$ be as introduced in Section 9.4. Let $\tilde{v}$, $\tilde{w}$ be fuzzy numbers representing the aspiration levels of Player I and Player II respectively. Then a *bi-matrix game with fuzzy goals and fuzzy payoffs*, denoted by $BGFGFP$, is defined as:

$$BGFGFP = (S^m,\ S^n,\ \tilde{A},\ \tilde{B},\ \tilde{v},\ \tilde{p},\ \tilde{p}',\ \tilde{w},\ \tilde{q},\ \tilde{q}',\ \lesssim,\ \gtrsim),$$

where, '$\gtrsim$' and '$\lesssim$' have their meanings as explained in Section 9.3 and, $\tilde{p}$, $\tilde{p}'$, and, $\tilde{q}'$, $\tilde{q}$ are fuzzy tolerance levels for Player I and Player II respectively.

We now define the meaning of an equilibrium solution of the fuzzy bi-matrix game $BGFGFP$.

Definition 9.5.1 (**Equilibrium solution of** $BGFGFP$). *A point* $(x^*,\ y^*) \in S^m \times S^n$ *is called an equilibrium solution to the fuzzy bi-matrix game* $BGFGFP$ *if*

$$\begin{aligned} x^T\tilde{A}y^* &\lesssim_{\tilde{p}} \tilde{v},\ \forall\ x \in S^m, \\ x^{*T}\tilde{B}y &\lesssim_{\tilde{q}} \tilde{w},\ \forall\ y \in S^n, \\ x^{*T}\tilde{A}y^* &\gtrsim_{\tilde{p}'} \tilde{v}, \\ x^{*T}\tilde{B}y^* &\gtrsim_{\tilde{q}'} \tilde{w}. \end{aligned}$$

By using the above definition for the game $BGFGFP$, we construct the following fuzzy non-linear programming problem ($FNP4$) for Player I and Player II

($FNP4$) Find $(x,\ y) \in R^m \times R^n$ such that

$$\begin{aligned} x^T\tilde{A}y &\lesssim_{\tilde{p}} \tilde{v},\ \forall\ x \in S^m, \\ x^T\tilde{B}y &\lesssim_{\tilde{q}} \tilde{w},\ \forall\ y \in S^n, \\ x^T\tilde{A}y &\gtrsim_{\tilde{p}'} \tilde{v}, \\ x^T\tilde{B}y &\gtrsim_{\tilde{q}'} \tilde{w}, \\ x &\in S^m, \\ y &\in S^n. \end{aligned}$$

Now employing the resolution method of Yager [87] for the double fuzzy constraints (as discussed here in Section 9.4) and following

Zimmermann's approach [90], the above fuzzy non-linear programming problems (*FNP*4) reduces to

(*FNP*5) max λ

subject to,

$$
\begin{aligned}
&x^T\tilde{A}y \lessapprox \tilde{v} + \tilde{p}(1-\lambda),\ \forall\ x \in S^m,\\
&x^T\tilde{B}y \lessapprox \tilde{w} + \tilde{q}(1-\lambda),\ \forall\ y \in S^n,\\
&x^T\tilde{A}y \gtrapprox \tilde{v} - \tilde{p}'(1-\lambda),\\
&x^T\tilde{B}y \gtrapprox \tilde{w} - \tilde{q}'(1-\lambda),\\
&\quad x \in S^m,\\
&\quad y \in S^n.
\end{aligned}
$$

Next, we utilize the defuzzification function $F : N(\mathbb{R}) \rightarrow R$ for the constraints of (*FNP*5), to get

(*NLP*4) max λ

subject to,

$$
\begin{aligned}
&F(x^T\tilde{A}y) \le F(\tilde{v}) + F(\tilde{p})(1-\lambda),\ \forall\ x \in S^m,\\
&F(x^T\tilde{B}y) \le F(\tilde{w}) + F(\tilde{q})(1-\lambda),\ \forall\ y \in S^n,\\
&F(x^T\tilde{A}y) \ge F(\tilde{v}) - F(\tilde{p}')(1-\lambda),\\
&F(x^T\tilde{B}y) \ge F(\tilde{w}) - F(\tilde{q}')(1-\lambda),\\
&\quad x \in S^m,\\
&\quad y \in S^n.
\end{aligned}
$$

As we have mentioned earlier, the defuzzification function F preserves the ranking when fuzzy numbers are multiplied by non-negative scalars, and therefore problem (*NLP*4) becomes

(*NLP*5) max λ

subject to,

$$
\begin{aligned}
&x^TF(\tilde{A})y \le F(\tilde{v}) + F(\tilde{p})(1-\lambda),\ \forall\ x \in S^m,\\
&x^TF(\tilde{B})y \le F(\tilde{w}) + F(\tilde{q})(1-\lambda),\ \forall\ y \in S^n,\\
&x^TF(\tilde{A})y \ge F(\tilde{v}) - F(\tilde{p}')(1-\lambda),\\
&x^TF(\tilde{B})y \ge F(\tilde{w}) - F(\tilde{q}')(1-\lambda),\\
&\quad x \in S^m,\\
&\quad y \in S^n.
\end{aligned}
$$

Since S^m and S^n are convex polytopes, it is sufficient to consider only the extreme points (i.e. pure strategies) of S^m and S^n in the constraints of (*NLP*5). This observation leads to the following non-linear programming problem, for Player I and Player II

(*NLP*6) max λ
subject to,

$$
\begin{aligned}
F(\tilde{A})_i y &\leq F(\tilde{v}) + F(\tilde{p})(1-\lambda), \quad (i = 1,2,\ldots,m),\\
x^T F(\tilde{B})_j &\leq F(\tilde{w}) + F(\tilde{q})(1-\lambda), \quad (j = 1,2,\ldots,n),\\
x^T F(\tilde{A}) y &\geq F(\tilde{v}) - F(\tilde{p}')(1-\lambda),\\
x^T F(\tilde{B}) y &\geq F(\tilde{w}) - F(\tilde{q}')(1-\lambda),\\
x &\in S^m,\\
y &\in S^n,\\
\lambda &\in [0,1].
\end{aligned}
$$

Here $F(\tilde{A})_i$ denotes the ith row of $F(\tilde{A})$ and $F(\tilde{B})_j$ denotes the jth column of $F(\tilde{B})$ where $i = 1,2,\ldots,m$, and $j = 1,2,\ldots,n$.

From the above discussion we now observe that for solving the fuzzy bi-matrix game *BGFGFP* we have to solve the crisp non-linear programming problem (*NLP*6). Also, if $(x^*,\ y^*,\ \lambda^*)$ is an optimal solution of (*NLP*6) then $(x^*,\ y^*)$ is an equilibrium solution for the game *BGFGFP* and λ^* is the degree to which the aspiration levels $F(\tilde{v})$ and $F(\tilde{w})$ of Player I and Player II can be met.

All the results discussed in this section can now be summarized in the form of Theorem 9.5.1 given below.

Theorem 9.5.1 *The fuzzy bi-matrix game BGFGFP described by*

$$BGFGFP = (S^m,\ S^n,\ \tilde{A},\ \tilde{B},\ \tilde{v},\ \tilde{p},\ \tilde{p}',\ \tilde{w},\ \tilde{q},\ \tilde{q}',\ \lesssim,\ \gtrsim)$$

is equivalent to the crisp non-linear programming problem (*NLP*6).

Remark 9.5.1. The (crisp) equivalents of fuzzy nonlinear programming problems (*FNP*), (*FNP*1) and (*FNP*4), namely (*NLP*), (*NLP*1) and (*NLP*6) have a special structure. Specifically, the problems (*NLP*), (*NLP*1) and (*NLP*6) are linear expect for two constraints which are quadratic in nature. This observation can possibly be exploited to develop satisfactory solution procedures for solving these structured nonlinear programming problems. In this context, it may be noted that there already exists an efficient LP based solution procedure to solve linear programming problems with one quadratic constraint, e.g. Van De Panne [73].

9.6 Bi-matrix game with fuzzy pay-offs: A possibility measure approach

In this section, we present Maeda's [49] possibility measure approach to characterize an equilibrium solution of a bi-matrix game with fuzzy pay-offs. Let us recall that if $\tilde{a}$ and $\tilde{b}$ are two fuzzy numbers then $\texttt{Poss}\ (\tilde{a} \geq \tilde{b})$ is defined as

$$\texttt{Poss}\ (\tilde{a} \geq \tilde{b}) = \sup_{x \geq y} \Big(\min \big(\mu_{\tilde{a}}(x),\ \mu_{\tilde{b}}(y) \big) \Big).$$

Also, if $\tilde{a}$ and $\tilde{b}$ are symmetric TFNs and $\alpha \in (0,1]$ then $\texttt{Poss}\ (\tilde{a} \geq \tilde{b}) \geq \alpha$ if and only if $a^R_\alpha \geq b^L_\alpha$, where $[a^L_\alpha,\ a^R_\alpha]$ and $[b^L_\alpha,\ b^R_\alpha]$ are α-cuts of $\tilde{a}$ and $\tilde{b}$ respectively. Similarly $\texttt{Poss}\ (\tilde{a} \geq \tilde{b}) \leq \alpha$ if $a^R_\alpha \leq b^L_\alpha$. This later result follows because $\texttt{Necc}\ (\tilde{a} > \tilde{b}) \geq \alpha$ if and only if $a^L_{1-\alpha} \geq b^R_{1-\alpha}$ and $\texttt{Poss}\ (\tilde{a} \geq \tilde{b}) = 1 - \texttt{Necc}\ (\tilde{b} > \tilde{a})$. Although Maeda's [49] presentation is valid in somewhat more generality (i.e. L fuzzy numbers) we have deliberately taken the fuzzy numbers to be symmetric TFNs.

We now consider the bi-matrix game with fuzzy pay-offs, denoted by *BFP*, and define its equilibrium solution in the sense of a possibility measure. Specifically, let $BFP = (S^m,\ S^n,\ \tilde{A},\ \tilde{B})$ where the elements $\tilde{a_{ij}}$ and $\tilde{b_{ij}}$ of pay-off matrices A and B respectively, are symmetric TFN's. Also, if the need be, let a real number v be identified as a TFN $\tilde{v} = (v,\ v,\ v)$. Therefore we shall be writing the real number v and also its TFN equivalent $\tilde{v} = (v,\ v,\ v)$ with out causing any confusion as the meaning will be clear from the given context.

Definition 9.6.1 (A $(v,\ w)$-possible Nash equilibrium strategy). *Let $v, w \in \mathbb{R}$. An element $(\bar{x},\ \bar{y}) \in S^m \times S^n$ is called a $(v,\ w)$-possible Nash equilibrium strategy if*

$$\texttt{Poss}\ (\bar{x}^T \tilde{A} \bar{y} \geq \tilde{v}) \geq \texttt{Poss}\ (x^T \tilde{A} \bar{y} \geq \tilde{v}),\ \forall x \in S^m,$$

and

$$\texttt{Poss}\ (\bar{x}^T \tilde{B} \bar{y} \geq \tilde{w}) \geq \texttt{Poss}\ (\bar{x}^T \tilde{B} y \geq \tilde{w}),\ \forall y \in S^n.$$

Here as agreed, $\tilde{v} = (v,\ v,\ v)$ and $\tilde{w} = (w,\ w,\ w)$.

It may now be observed that for given $v,\ w \in \mathbb{R}$, any $\bar{x} \in S^m$, $\bar{y} \in S^n$ such that $\texttt{Poss}\ (\bar{x}^T \tilde{A} \bar{y} \geq \tilde{v}) = 1 = \texttt{Poss}\ (\bar{x}^T \tilde{B} \bar{y} \geq \tilde{w})$, is certainly a (v, w)-possible Nash equilibrium strategy.

Definition 9.6.2 (An (α, β)-Nash equilibrium strategy). *Let $\alpha, \beta \in [0,1]$. An element $(\bar{x}, \bar{y}) \in S^m \times S^n$ is said to be an (α, β)-Nash equilibrium strategy of BFP if*

$$\bar{x}^T A^R_\alpha \bar{y} \geq x^T A^R_\alpha \bar{y}, \ \forall x \in S^m,$$

and

$$\bar{x}^T B^R_\beta \bar{y} \geq \bar{x}^T B^R_\beta y, \ \forall y \in S^n.$$

Here A^R_α is a matrix whose entries are $(\tilde{a}_{ij})^R_\alpha$ and $[(\tilde{a}_{ij})^R_\alpha, (\tilde{a}_{ij})^L_\alpha]$ is the α-cut of $\tilde{a}_{ij}$. The matrix B^R_β is defined similarly.

For the fuzzy bi-matrix game *BFP*, we have the following theorem which connects a (v, w)-possible Nash equilibrium strategy with (α, β)-Nash equilibrium strategy.

Theorem 9.6.1 *Let $v, w \in \mathbb{R}$ and $(\bar{x}, \bar{y}) \in S^m \times S^n$ be any (v, w)-possible Nash equilibrium strategy for the fuzzy matrix game BFP. Let $\alpha = \text{Poss}\,(\bar{x}^T \tilde{A} \bar{y} \geq \tilde{v})$, $\beta = \text{Poss}\,(\bar{x}^T \tilde{B} \bar{y} \geq \tilde{w})$ and $\alpha, \beta \in (0,1)$. Then $(\bar{x}, \bar{y})$ is also an (α, β)-Nash equilibrium strategy. In that case $v = \bar{x}^T A^R_\alpha \bar{y}$ and $w = \bar{x}^T B^R_\beta \bar{y}$.*

Conversely, for any $\alpha, \beta \in [0,1]$, let $(\bar{x}, \bar{y})$ be an (α, β)-Nash equilibrium strategy of the game BFP. Then $(\bar{x}, \bar{y})$ is also a (v, w)-possible Nash equilibrium strategy where $v = \bar{x}^T A^R_\alpha \bar{y}$, $w = \bar{x}^T B^R_\beta \bar{y}$.

Proof. Given that $(\bar{x}, \bar{y}) \in S^m \times S^n$ is a (v, w)-possible Nash equilibrium strategy for *BFP*, we have

$$\text{Poss}\,(\bar{x}^T \tilde{A} \bar{y} \geq \tilde{v}) \geq \text{Poss}\,(x^T \tilde{A} \bar{y} \geq \tilde{v}), \ \forall x \in S^m,$$

and

$$\text{Poss}\,(\bar{x}^T \tilde{B} \bar{y} \geq \tilde{w}) \geq \text{Poss}\,(\bar{x}^T \tilde{B} y \geq \tilde{w}), \ \forall y \in S^n.$$

Now setting,

$$\alpha = \text{Poss}\,(\bar{x}^T \tilde{A} \bar{y} \geq \tilde{v}),$$

and

$$\beta = \text{Poss}\,(\bar{x}^T \tilde{B} \bar{y} \geq \tilde{w}),$$

the above inequations imply

$$\text{Poss}\,(x^T \tilde{A} \bar{y} \geq \tilde{v}) \leq \alpha, \ \forall x \in S^m,$$

and

$$\text{Poss}\,(\bar{x}^T\tilde{B}y \geq \tilde{w}) \leq \beta,\ \forall y \in S^n,$$

which by the definition of possibility for the case of symmetric TFNs means

$$x^T A^R_\alpha \bar{y} \leq v^L_\alpha,\ \forall x \in S^m,$$

and

$$\bar{x}^T B^R_\beta y \leq w^L_\beta,\ \forall y \in S^n.$$

But $\text{Poss}\,(\bar{x}^T\tilde{A}\ \bar{y} \geq \tilde{v}) = \alpha$ means $\bar{x}^T A^R_\alpha \bar{y} = v^L_\alpha = v$ as $\tilde{v} = (v,\ v,\ v)$. Therefore

$$x^T A^R_\alpha \bar{y} \leq \bar{x}^T A^R_\alpha \bar{y},\ \forall x \in S^m.$$

Similarly, $\bar{x}^T B^R_\beta \bar{y} = w^L_\beta = w$ and

$$\bar{x}^T B^R_\beta y \leq \bar{x}^T B^R_\beta \bar{y},\ \forall y \in S^n,$$

implying that $(\bar{x},\ \bar{y})$ is an $(\alpha,\ \beta)$-Nash equilibrium strategy of the fuzzy game *BFP*.
Conversely let $(\bar{x},\ \bar{y})$ be an $(\alpha,\ \beta)$-Nash equilibrium strategy. We have to show that $(\bar{x},\ \bar{y})$ is also a $(v,\ w)$-possible Nash equilibrium strategy where $v = \bar{x}^T A^R_\alpha\ \bar{y}$ and $w = \bar{x}^T B^R_\beta\ \bar{y}$.
Now for $\alpha = \beta = 1$, it is obvious that $(\bar{x},\ \bar{y})$ is a $(v,\ w)$-possible Nash equilibrium strategy. So without any loss of generality we assume that $\alpha < 1$. If possible, let there exist $x \in S^m$ such that $\text{Poss}\,(x^T\tilde{A}\bar{y} \geq \tilde{v}) = \bar{\alpha} > \alpha$ holds. Then we have

$$x^T A^R_\alpha\ \bar{y} > x^T A^R_{\bar{\alpha}}\ \bar{y} \geq v^L_\alpha = v = \bar{x}^T A^R_\alpha\ \bar{y},$$

which is a contradiction. Therefore

$$\text{Poss}\,(\bar{x}^T\tilde{A}\bar{y} \geq \tilde{v}) \geq \text{Poss}\,(x^T\tilde{A}\bar{y} \geq \tilde{v}),\ \forall x \in S^m.$$

Similarly,

$$\text{Poss}\,(\bar{x}^T\tilde{B}\bar{y} \geq \tilde{w}) \geq \text{Poss}\,(\bar{x}^T\tilde{B}y \geq \tilde{w}),\ \forall y \in S^n.$$

Therefore $(\bar{x},\ \bar{y})$ is a $(v,\ w)$-possible Nash equilibrium strategy.

We now have the following results on the existence of Nash equilibrium strategies for the fuzzy bi-matrix game *BFP*.

Theorem 9.6.2 *Every fuzzy bi-matrix game BFP has a* $(\alpha,\ \beta)$*-Nash equilibrium strategy* $(\bar{x},\ \bar{y})$ *for every* $\alpha, \beta \in [0, 1]$.

Theorem 9.6.3 *Let $\alpha, \beta \in [0,1]$. Then a point $(\bar{x}, \bar{y}) \in S^m \times S^n$ is an (α, β)-Nash equilibrium strategy of BFP if and only if $(\bar{x}, \bar{y}, \bar{v} = \bar{x}^T A^R_\alpha \bar{y}, \bar{w} = \bar{x}^T B^R_\beta \bar{y})$ is an optimal solution of the following quadratic programming problem*

$$\begin{aligned} \max \quad & x^T(A^R_\alpha + B^R_\alpha)y - v - w \\ \text{subject to,} \quad & \\ & A^R_\alpha y \le ev, \\ & (B^R_\beta)^T x \le ew, \\ & e^T x = 1, \\ & e^T y = 1, \\ & x, y \ge 0, \\ & v, w \in \mathbb{R}. \end{aligned}$$

The proofs of above two theorems follow from Definition 9.6.2, Theorem 1.5.1 and Theorem 1.6.1 of Chapter 1.

The next theorem in this sequel states that the fuzzy bi-matrix game *BFP* has at least one (v, w)-possible Nash equilibrium strategy for every $v, w \in \mathbb{R}$.

Theorem 9.6.4 *For every $v, w \in \mathbb{R}$ the fuzzy bi-matrix game BFP has at least one (v, w)-Nash equilibrium strategy.*

Proof. The proof of the above theorem follows similar to that of the crisp case. For this we consider the problem

$$\begin{aligned} (FLP)_{(v,w)} \qquad \max \quad & \text{Poss}\,(x^T \tilde{A} \bar{y} \ge \tilde{v}) + \text{Poss}\,(\bar{x}^T \tilde{B} y \ge \tilde{w}) \\ \text{subject to,} \quad & \\ & (x, y) \in S^m \times S^n, \end{aligned}$$

and denote by $F(\bar{x}, \bar{y})$ as the set of all optimal solutions of $(FLP)_{(v,w)}$. It can be shown that the set valued map $F : S^m \times S^n \to S^m \times S^n$ is upper semi-continuous, closed and convex and therefore, by Kakutani's fixed point theorem has a fixed point i.e. there exists $(\bar{x}, \bar{y}) \in S^m \times S^n$ such that $(\bar{x}, \bar{y}) \in F(\bar{x}, \bar{y})$. The rest of the proof is now similar to Theorem 1.5.1.

Now instead of problem $(FLP)_{(v,w)}$, it is convenient to consider the following problem $P_{(v,w)}$:

$$\begin{aligned} (P)_{(v,w)} \qquad \max \quad & \text{Poss}\,(x^T \tilde{A} \bar{y} \ge \tilde{v}) + \text{Poss}\,(\bar{x}^T \tilde{B} y \ge \tilde{w}) - \alpha - \beta \\ \text{subject to,} \quad & \end{aligned}$$

$$\begin{aligned} \text{Poss}\,(\tilde{A}y \geq e\tilde{v}) &\leq \alpha, \\ \text{Poss}\,(\tilde{B}^T x \geq e\tilde{w}) &\leq \beta, \\ e^T x &= 1, \\ e^T y &= 1, \\ x, y &\geq 0, \\ \alpha, \beta &\in [0,1]. \end{aligned}$$

The following theorem seems to be very natural in this context.

Theorem 9.6.5 *Let $v, w \in \mathbb{R}$ and a point $(\bar{x}, \bar{y}) \in S^m \times S^n$ be a (v, w)-possible Nash equilibrium strategy for the game BFP. Then $(\bar{x}, \bar{y}, \bar{\alpha}, \bar{\beta})$ is an optimal solution of $P_{(v, w)}$, where $\bar{\alpha} = \text{Poss}\,(\bar{x}^T \tilde{A}\bar{y} \geq \tilde{v})$ and $\bar{\beta} = \text{Poss}\,(\bar{x}^T \tilde{B}\bar{y} \geq \tilde{w})$.*
Conversely, let $(\bar{x}, \bar{y}, \bar{\alpha}, \bar{\beta})$ be an optimal solution of the problem $P_{(v, w)}$. Then $(\bar{x}, \bar{y})$ is a (v, w)-possible Nash equilibrium strategy of BFP. Also,

$$\bar{\alpha} = \text{Poss}\,(\bar{x}^T \tilde{A}\bar{y} \geq \tilde{v}),$$

and

$$\bar{\beta} = \text{Poss}\,(\bar{x}^T \tilde{B}\bar{y} \geq \tilde{w}).$$

Proof. Let us first note that for any feasible solution (x, y, α, β) of the problem

$(P)_{(v, w)}$ $\quad \text{Poss}\,(x^T \tilde{A}y \geq \tilde{v}) + \text{Poss}\,(x^T \tilde{B}y \geq \tilde{w}) - \alpha - \beta \leq 0.$

We now prove necessary part. For this let $(\bar{x}, \bar{y}) \in S^m \times S^n$ be a (v, w)-possible Nash equilibrium strategy of the game BFP. Also let $\bar{\alpha} = \text{Poss}\,(\bar{x}^T \tilde{A}\bar{y} \geq \tilde{v})$ and $\bar{\beta} = \text{Poss}\,(\bar{x}^T \tilde{B}\bar{y} \geq \tilde{w})$. From the fact that for any feasible solution of $P_{(v, w)}$ the objective function value is less than or equal to zero, it is enough to prove that $(\bar{x}, \bar{y}, \bar{\alpha}, \bar{\beta})$ is feasible to $P_{(v, w)}$. But this is obvious for the case $\bar{\alpha} = \bar{\beta} = 1$. Therefore without any loss of generality let $\bar{\alpha}, \bar{\beta} \in [0, 1)$.

Then as $(\bar{x}, \bar{y})$ is a (v, w)-Nash equilibrium strategy and $\bar{\alpha} = \text{Poss}\,(\bar{x}^T \tilde{A}\bar{y} \geq \tilde{v})$, $\bar{\beta} = \text{Poss}\,(\bar{x}^T \tilde{B}\bar{y} \geq \tilde{w})$, we have by definition that

$$x^T A^R_{\bar{\alpha}}\, \bar{y} \leq v, \ \forall x \in S^m,$$

and

$$\bar{x}^T B^R_{\bar{\beta}}\, y \leq w, \ \forall y \in S^n.$$

Therefore $A^R_{\alpha}\bar{y} \leq ev$ and $(B^R_{\beta})^T \bar{x} \leq ew$ and hence $\text{Poss}\,(\tilde{A}\bar{y} \geq \tilde{v}e) \leq \bar{\alpha}$ and $\text{Poss}\,(\tilde{B}^T \bar{x} \geq \tilde{w}e) \leq \bar{\beta}$. This gives that $(\bar{x}, \bar{y}, \bar{\alpha}, \bar{\beta})$ is feasible to $P_{(v, w)}$.
Conversely, let $(\bar{x}, \bar{y}, \bar{\alpha}, \bar{\beta})$ be an optimal solution of $P_{(v, w)}$. We shall

show that $(\bar{x}, \bar{y})$ is a (v, w)-Nash equilibrium strategy of the fuzzy matrix game BFP. The optimality of $(\bar{x}, \bar{y}, \bar{\alpha}, \bar{\beta})$ to $P_{(v,w)}$ gives Poss $(\bar{x}^T\tilde{A}\bar{y} \geq \tilde{v}) = \bar{\alpha}$ and Poss $(\bar{x}^T\tilde{B}\bar{y} \geq \tilde{w}) = \bar{\beta}$. In case $\bar{\alpha} = \bar{\beta} = 1$, the proof is obvious. So let us now assume that $\bar{\alpha}, \bar{\beta} \in [0,1)$. Then, the feasibility of $(\bar{x}, \bar{y}, \bar{\alpha}, \bar{\beta})$ to $P_{(v,w)}$ gives Poss $(\tilde{A}\bar{y} \geq e\tilde{v}) \leq \bar{\alpha}$ and Poss $(\tilde{B}^T\bar{x} \geq e\tilde{w}) \leq \bar{\beta}$. But these imply $A^R_{\bar{\alpha}}\,\bar{y} \leq ev$ and $(B^R_{\bar{\beta}})^T\,\bar{x} \leq ew$, i.e.

$$x^T A^R_{\bar{\alpha}}\,\bar{y} \leq v, \;\forall x \in S^m,$$

and

$$\bar{x}^T B^R_{\bar{\beta}}\,y \leq w, \;\forall y \in S^n.$$

From the above inequalities it follows by definition, that

$$\text{Poss}\,(x^T\tilde{A}\bar{y} \geq \tilde{v}) \leq \bar{\alpha} = \text{Poss}\,(\bar{x}^T\tilde{A}\bar{y} \geq \tilde{v}),\; \forall x \in S^m,$$

and

$$\text{Poss}\,(\bar{x}^T\tilde{B}y \geq \tilde{w}) \leq \bar{\beta} = \text{Poss}\,(\bar{x}^T\tilde{B}\bar{y} \geq \tilde{w}),\; \forall y \in S^n.$$

This proves that $(\bar{x}, \bar{y})$ is a (v, w)-possible Nash equilibrium strategy of the game BFP.

9.7 Conclusions

In this chapter we have studied fuzzy bi-matrix games having fuzzy goals or fuzzy pay-offs or both. While for the case when only goals are fuzzy, the main model presented is due to Nishizaki and Sakawa [61] which is very much in the spirit of Mangasarian and Stone [52] for the crisp case. For the case when pay-offs are fuzzy or goals and pay-offs both are fuzzy, a ranking function approach is developed to solve such fuzzy bi-matrix games. This approach is different from that of Nishizaki and Sakawa [61] and seems to be simple provided we can choose a ranking function appropriately. We have not discussed Nishizaki and Sakawa's model for the general (when goals as well as pay-offs are fuzzy) fuzzy scenario where a different and seemingly more difficult mathematical programming problem needs to be solved. Nevertheless, it is an important development and for that we need to refer to Nishizaki and Sakawa [61]. Another direction in which fuzzy bi-matrix games have been studied is the Maeda's [49] possibility measure approach which we have discussed here in Section 9.6. A close look at the Maeda's formulation suggests that this approach could possibility be investigated for fuzzy matrix games as well and results could be related to fuzzy linear programming duality in a possibility measure setting.

10

Modality and other approaches for fuzzy linear programming

10.1 Introduction

Fuzzy Mathematical programming problems can be classified on the basis of two concepts. The first concept is the fuzziness of the decision maker's aspirations with respect to goals and/or constraints. The other one is the ambiguity of the coefficients of the objective function and/or constraints. Combination of these two concepts gives us different types of fuzzy mathematical programming problems and some of these have been studied in Chapter 4.

In this chapter we follow a more general approach of introducing a broad class of fuzzy linear programming problems, which embeds the classical (crisp) linear programming problems as well as various other fuzzy linear programming problems discussed in Chapter 4. This chapter is divided in seven main sections, namely, *fuzzy measure*, *fuzzy preference relations*, *modality constrained programming problems*, *valued relations and their fuzzy extensions*, *fuzzy linear programming via fuzzy relations*, *duality in fuzzy linear programming via fuzzy relations*, *duality in fuzzy linear programming with fuzzy coefficients: Wu's model.*

10.2 Fuzzy measure

A fuzzy measure describes the uncertainty or imprecision in assigning a particular element to certain given crisp subsets. In order to describe this type of uncertainty, a value is assigned to each of these crisp subsets. This value signifies the degree of evidence or belief of that particular element's membership in the set. Such type of representation of uncertainty is known as a *fuzzy measure.* There is a conceptual difference between a fuzzy measure and a fuzzy set. In the case of a fuzzy set, a value is assigned to each element of the universal set signifying its degree of membership in a particular set whose boundaries are not sharp, where as in the case of a fuzzy measure, the sets are crisp and therefore have well defined boundaries. Here the value assigned to each crisp set of the universal set represents the degree of evidence or belief that a particular element, which is not fully characterized, is in the set. We formally define a fuzzy measure as follows:

Definition 10.2.1 (Fuzzy measure). *Let X be a universal set and $\mathsf{P}(X)$ be the power set of X. Then a fuzzy measure on $\mathsf{P}(X)$ is a function $g : \mathsf{P}(X) \to [0,1]$ satisfying the following:*

(i) $g(\phi) = 0$ and $g(X) = 1$ (boundary conditions),
(ii) for any A, $B \in \mathsf{P}(X)$, if $A \subseteq B$ then $g(A) \leq g(B)$ (monotonicity),
(iii) for any increasing sequence $A_1 \subseteq A_2 \subseteq \ldots$ in $\mathsf{P}(x)$,

$$\lim_{i\to\infty} g(A_i) = g\left(\bigcup_{i=1}^{\infty} A_i\right)$$

(continuity from below),
(iv) for any decreasing sequence $A_1 \supseteq A_2 \supseteq \ldots$ in $\mathsf{P}(x)$,

$$\lim_{i\to\infty} g(A_i) = g\left(\bigcap_{i=1}^{\infty} A_i\right)$$

(continuity from above).

This function g assigns a number $g(A)$ in the unit interval $[0,1]$ to each crisp subset A of X.

The number $g(A)$ represents the degree of the available evidence or our belief that a given element of X, which is not fully characterized, belongs to the subset A. A typical example could be the classification of an item as "satisfactory" or "unsatisfactory" for a job. Suppose it requires a large number of cumbersome measurements to classify this item exactly in one of these two classes, but by having some simple preliminary measurements we may be able to have a guess that the item in question belongs to the class "satisfactory" with the belief 0.75 and it belongs to the class "unsatisfactorily" with the belief 0.45. Then if S denotes the set of all satisfactory items and U denotes the set of all unsatisfactorily items, then $g(S) = 0.75$ and $g(U) = 0.45$ define a fuzzy measure.

From the above definition it is observed that

$$g(A \cup B) \geq \max\,(g(A), g(B))$$

and

$$g(A \cap B) \leq \min\,(g(A), g(B)).$$

This is because

$$(A \cap B) \subseteq A \subseteq (A \cup B),\ (A \cap B) \subseteq B \subseteq (A \cup B)$$

and g is monotonic.

There are three special forms of fuzzy measures; namely evidence theory, possibility theory and probability theory, out of which we discuss here only the first two.

Evidence theory

This theory explains two dual measures i.e. belief measures and plausibility measures which are defined below.

Definition 10.2.2 (Belief measure). *A belief measure is a function* $\mathtt{bel} : \mathsf{P}(X) \to [0,1]$ *such that* $\mathtt{bel}$ *is a fuzzy measure and*

$$\mathtt{bel}\left(\bigcup_{j=1}^{n} A_j\right) \geq \sum_j \mathtt{bel}(A_j) - \sum_{j<k} \mathtt{bel}\left(A_j \bigcap A_k\right) + \ldots + (-1)^{n+1}\mathtt{bel}\left(\bigcap_{j=1}^{n} A_j\right)$$

where $A_1, A_2, \ldots, A_n$ *are* n *crisp subsets of the universe* X.

Definition 10.2.3 (Plausibility measure). *A plausibility measure is a mapping* $\mathtt{pl} : \mathsf{P}(X) \to [0,1]$ *such that* $\mathtt{pl}$ *is a fuzzy measure and*

$$\mathtt{pl}\left(\bigcap_{j=1}^{n} A_j\right) \geq \sum_j \mathtt{pl}\ (A_j) - \sum_{j<k} \mathtt{pl}\ \left(A_j \bigcup A_k\right) + \ldots + (-1)^{n+1}\mathtt{pl}\ \left(\bigcup_{j=1}^{n} A_j\right)$$

where $A_1, A_2, \ldots, A_n$ *are* n *crisp subsets of the universe* X.

We can verify the following from the above definitions:

(i) $\mathtt{bel}\ (A) = 1 - \mathtt{pl}\ (\bar{A})$, $\mathtt{pl}\ (A) = 1 - \mathtt{bel}\ (\bar{A})$.
(ii) $\mathtt{bel}\ (A) + \mathtt{bel}\ (\bar{A}) \leq 1$, $\mathtt{pl}\ (A) + \mathtt{pl}\ (\bar{A}) \geq 1$.

(iii) $\texttt{bel}\ (A) \leq \texttt{pl}\ (A)$.

The above results can be given meaningful interpretations. For example, $\texttt{bel}\ (A) + \texttt{bel}\ (\bar{A}) \leq 1$ indicates that the lack of belief in $x \in A$ does not mean a strong belief in $x \in \bar{A}$.

Now both belief and plausibility measures can be characterized by only one function namely $m : \mathsf{P}(X) \to [0,1]$ with $m(\phi) = 0$ and $\sum_{A \in \mathsf{P}(X)} m(A) = 1$. This function is called *basic probability assignment.* The value $m(A)$, for $A \in \mathsf{P}(X)$ is called A's *basic probability number* and is considered as the degree of evidence indicating that a specific element of X belongs to the set A alone but not to any special subset of A.

Here it may be noted that a basic probability assignment m is different from a probability distribution function because a probability distribution function is defined on X where as m is defined on $\mathsf{P}(X)$. Also, m is not necessarily a fuzzy measure because we do not require that $m(X) = 1$ or that it is monotonic.

Now given a basic assignment m, a belief measure and a plausibility measure can uniquely be determined by

$$\texttt{bel}\ (A) = \sum_{B \subseteq A} m(B),$$

and

$$\texttt{pl}\ (A) = \sum_{B \cap A \neq \phi} m(B),$$

for all $A \in \mathsf{P}(X)$.

The relationship between $m(A)$, $\texttt{bel}\ (A)$ and $\texttt{pl}\ (A)$ can now be summarized as

(i) $m(A)$ measures the belief that the element in question, namely $x \in X$ belongs to the set A alone.
(ii) $\texttt{bel}\ (A)$ gives the total evidence or belief that the element $x \in X$ belongs to the set A and to the various special subsets of A.
(iii) $\texttt{pl}\ (A)$ represents the total evidence or belief that the element $x \in X$ belongs to the set A or to any of the various special subsets of A together with the additional evidence or belief associated with the sets that overlap with A. This again explains the relation $\texttt{pl}\ (A) \geq \texttt{bel}\ (A)$.

Possibility theory

In evidence theory, we have two mutually dual measures i.e. belief and plausibility measures, where as in possibility theory, the counterparts of belief measure and plausibility measure are defined as necessity measure and possibility measure respectively. These are defined as follows:

Definition 10.2.4 (Necessity measure). *A necessity measure* Necc *is a fuzzy measure such that*

$$\texttt{Necc}\ (A \cap B) = \min(N(A), N(B)) \ \textit{for all}\ A, B \in \mathsf{P}(X).$$

Definition 10.2.5 (Possibility measure). *A possibility measure* Poss *is a fuzzy measure such that*

$$\texttt{Poss}\ (A \cup B) = \max(P(A), P(B)) \ \textit{for all}\ A, B \in \mathsf{P}(X).$$

On the basis of the previous results of plausibility measure and belief measure, the following results for possibility measure and necessity measure hold

$$\texttt{Necc}\ (A) + \texttt{Necc}\ (\bar{A}) \leq 1,$$

$$\texttt{Poss}\ (A) + \texttt{Poss}\ (\bar{A}) \geq 1,$$

$$\texttt{Necc}\ (A) = 1 - \texttt{Poss}\ (\bar{A}),$$

$$\texttt{Poss}\ (A) = 1 - \texttt{Necc}\ (\bar{A}),$$

$$\texttt{Necc}\ (A) > 0 \Leftrightarrow \texttt{Poss}\ (\bar{A}) = 1,$$

and

$$\texttt{Poss}\ (\bar{A}) < 1 \Leftrightarrow \texttt{Necc}\ (A) = 0.$$

For a given possibility measure Poss on $\mathsf{P}(X)$, a *possibility distribution function* associated with Poss is defined by the function $r : X \to [0,1]$ where $r(x) = \texttt{Poss}\ (\{x\})$ for all $x \in X$. Hence, every possibility measure is uniquely represented by the associated possibility distribution function. For finite universal sets, this property gives $\texttt{Poss}\ (A) = \max_{x \in A} r(x)$ for each $A \in \mathsf{P}(X)$.

In crisp scenario, let A and B be two sets and a variable x be in A. Now if $A \cap B \neq \phi$ holds then it can be said that $x \in B$ is possible under the information $x \in A$. But if $A \subseteq B$ holds, then it can be said that $x \in B$ is necessary under the information $x \in A$. Thus in the crisp scenario, we can define a possibility measure as follows:

$$P_A(B) = \begin{cases} 1, A \cap B \neq \phi, \\ 0, \text{otherwise.} \end{cases}$$

Similarly, a necessary measure can be defined as

$$N_A(B) = \begin{cases} 1, A \subseteq B, \\ 0, \text{otherwise.} \end{cases}$$

The above discussion motivates to define possibility and necessity measures for fuzzy sets A and B as follows:

$$P_A(B) = \sup_x \min \{\mu_A(x), \mu_B(x)\},$$

$$N_A(B) = \inf_x \max \{(1 - \mu_A(x)), \mu_B(x)\},$$

where μ_A and μ_B are membership functions of fuzzy sets A and B. Hereafter these indices, namely $P_A(B)$ and $N_A(B)$, will be called *modalities* because the degree of belonging of x to the fuzzy set B is considered from the modal point of view.

10.3 Fuzzy preference relations

Dubois and Prade [16] have used modalities to extend relations between real numbers to relations between fuzzy numbers. For this, let $\mathcal{R}$ be a relation between elements in X. Here we first study the extension of $\mathcal{R}$ to a relation between an element and a fuzzy set and then the extension of $\mathcal{R}$ to a relation between two fuzzy sets in the sense of Dubois and Prade [16].

Extension to relations between an element and a fuzzy set

These relations are as follows:

(i) The relation ${}_{P_l}\mathcal{R}$ with the membership function $\mu_{P_l\mathcal{R}}$, where x is possibly preferable to B, is given by:

$$\mu_{P_l\mathcal{R}}(x, B) = \sup_y \{\mu_B(y) \wedge \mu_{\mathcal{R}}(x, y)\}.$$

(ii) The relation ${}_{N_l}\mathcal{R}$ with the membership function $\mu_{N_l\mathcal{R}}$, where x is necessarily preferable to B, is given by:

$$\mu_{N_l\mathcal{R}}(x,B) = \inf_y \left\{\left(1-\mu_B(y)\right) \vee \mu_{\mathcal{R}}(x,y)\right\}.$$

(iii) The relation ${}_{P_f}\mathcal{R}$ with the membership function $\mu_{P_f\mathcal{R}}$, where B is possibly preferable to x, is given by:

$$\mu_{P_f\mathcal{R}}(B,x) = \sup_y \{\mu_B(y) \wedge \mu_{\mathcal{R}}(y,x)\}.$$

(iv) The relation ${}_{N_f}\mathcal{R}$ with the membership function $\mu_{N_f\mathcal{R}}$, where B is necessarily preferable to x, is given by:

$$\mu_{N_f\mathcal{R}}(B,x) = \inf_y \left\{\left(1-\mu_B(y)\right) \vee \mu_{\mathcal{R}}(y,x)\right\}.$$

Here the notations P_l (respectively N_l) and P_f (respectively N_f) means possibility (respectively necessity) operator with respect to latter and former arguments respectively to the ordered pairs.

Extension to relations between fuzzy sets

In order to extend the relation $\mathcal{R}$ to a relation between fuzzy sets, we use the above mentioned extended fuzzy preference relations and obtain

(i) The relation ${}_{P_fP_l}\mathcal{R}$ with the membership function $\mu_{P_fP_l\mathcal{R}}$, where it is possible for A to be possibly preferable to B; or, equivalently, the relation ${}_{P_lP_f}\mathcal{R}$ with the membership function $\mu_{P_lP_f\mathcal{R}}$ where it is possible for B to be a fuzzy set to which A is possibly preferable:

$$\mu_{P_fP_l\mathcal{R}}(A,B) = \sup_{x,y} \{\mu_A(x) \wedge \mu_B(y) \wedge \mu_{\mathcal{R}}(x,y)\} = \mu_{P_lP_f\mathcal{R}}(A,B).$$

(ii) The relation ${}_{P_fN_l}\mathcal{R}$ with the membership function $\mu_{P_fN_l\mathcal{R}}$, where it is possible for A to be necessarily preferable to B:

$$\mu_{P_fN_l\mathcal{R}}(A,B) = \sup_x \inf_y \left\{\mu_A(x) \wedge \left\{\left(1-\mu_B(y)\right) \vee \mu_{\mathcal{R}}(x,y)\right\}\right\}.$$

(iii) The relation ${}_{N_fP_l}\mathcal{R}$ with the membership function $\mu_{N_fP_l\mathcal{R}}$, where it is necessary for A to be possibly preferable to B:

$$\mu_{N_fP_l\mathcal{R}}(A,B) = \inf_x \sup_y \left\{\left(1-\mu_A(x)\right) \vee \{\mu_B(y)\right) \wedge \mu_{\mathcal{R}}(x,y)\}\right\}.$$

(iv) The relation ${}_{N_f N_l}\mathcal{R}$ with the membership function $\mu_{N_f N_l \mathcal{R}}$, where it is necessary for A to be necessarily preferable to B; or, equivalently, the relation ${}_{N_l N_f}\mathcal{R}$ with the membership function $\mu_{N_l N_f \mathcal{R}}$ where it is necessary for B to be a fuzzy set to which A is necessarily preferable:

$$\begin{aligned}\mu_{N_f N_l \mathcal{R}}(A,B) &= \inf_{x,y}\left\{\left(1-\mu_A(x)\right) \vee \left(1-\mu_B(y)\right) \vee \mu_{\mathcal{R}}(x,y)\right\} \\ &= \mu_{N_l N_f \mathcal{R}}(A,B).\end{aligned}$$

(v) The relation ${}_{N_l P_f}\mathcal{R}$ with the membership function $\mu_{N_l P_f \mathcal{R}}$, where it is necessary for B to be a fuzzy set to which A is possibly preferable:

$$\mu_{N_l P_f \mathcal{R}}(A,B) = \inf_y \sup_x \left\{\left(1-\mu_B(y)\right) \vee \{\mu_A(x)\right) \wedge \mu_{\mathcal{R}}(x,y)\}\right\}.$$

(vi) The relation ${}_{P_l N_f}\mathcal{R}$ with the membership function $\mu_{P_l N_f \mathcal{R}}$, where it is possible for B to be a fuzzy set to which A is necessarily preferable:

$$\mu_{P_l N_f \mathcal{R}}(A,B) = \sup_y \inf_x \left\{\mu_B(y) \wedge \left\{\left(1-\mu_A(x)\right) \vee \mu_{\mathcal{R}}(x,y)\right\}\right\}.$$

10.4 Modality constrained programming problems

In this section, we present a modality approach to study fuzzy mathematical programming problems having coefficients as fuzzy numbers (possibility distributions) and term the resulting problems as modality constrained programming problems. Most of the results on modality constrained programming are due to Inuiguchi et al. ([26], [27] and [28]) and this presentation is based on the work reported in [28]. A typical problem of this class could be described as

$$\begin{aligned}(FMP)\qquad &\text{optimize} \quad \Phi_1(x)\\ &\text{optimize} \quad \Phi_2(x)\\ &\qquad\qquad \vdots\\ &\text{optimize} \quad \Phi_p(x)\\ &\text{subject to,}\\ &\qquad F_i(x)\ \zeta_i\ G_i(x),\ i=1,2,\ldots,m,\\ &\qquad x \geq 0,\end{aligned}$$

where $x \in \mathbb{R}^n$, $\Phi_k(x) = \Phi_{k1}(x) \times \Phi_{k2}(x) \times \ldots \times \Phi_{kq}(x)$, $(k = 1,2,\ldots,p)$, $\Phi_{kj} : \mathbb{R}^n \to \mathbf{P}(\mathbb{R})$, $(i = 1,2,\ldots,p)$, $(j = 1,2,\ldots,q)$, $F_i(x) = F_{i1}(x) \times \ldots \times$

$F_{ir}(x)$, $G_i(x) = G_{i1}(x) \times \ldots \times G_{ir}(x), (i = 1, 2, \ldots, m)$, $F_{ij} : \mathbb{R}^n \to \mathbf{P}(\mathbb{R}), G_{ij} : \mathbb{R}^n \to \mathbf{P}(\mathbb{R}), (i = 1, 2, \ldots, m)$, $(j = 1, 2, \ldots, r_i)$, and $\mathbf{P}(\mathbb{R})$ is a set of possibility distributions in $\mathbb{R}$. Thus $\Phi_{ij}(x)$, $F_{ij}(x)$ and $G_{ij}(x)$ are possibility distribution in $\mathbb{R}$. Further, ζ_i, $(i = 1, 2, \ldots, m)$ are binary relations in $\mathbb{R}^{r_i}$ which are assumed as fuzzy preference relations.

In (*FMP*) if there is no objective function then it will be called a *system of fuzzy constraints.* Also, if $p > 1$ then it is called a *fuzzy multiobjective programming problem.*

To formulate the fuzzy mathematical programming problem (*FMP*), we first present a method to handle a constraint of type $F(x)\ \zeta\ G(x)$, where $F(x)$ and $G(x)$ both are possibility distributions and the relation ζ is a fuzzy preference relation. This type of constraints are called *preference constraints.*

In order to simplify the preference constraint $F(x)\zeta\ G(x)$, the fuzzy preference relation ζ between the possibility distributions can be understood by modalities suggested by Dubois and Prade as given in Section 10.3. Therefore, we get following modality constraints:

$$\mu_{P_f P_l}\zeta(F(x), G(x)) \geq \lambda_1,$$

$$\mu_{P_f N_l}\zeta(F(x), G(x)) \geq \lambda_2,$$

$$\mu_{N_f P_l}\zeta(F(x), G(x)) \geq \lambda_3,$$

$$\mu_{N_f N_l}\zeta(F(x), G(x)) \geq \lambda_4,$$

$$\mu_{N_l P_f}\zeta(F(x), G(x)) \geq \lambda_5,$$

$$\mu_{P_l N_f}\zeta(F(x), G(x)) \geq \lambda_6,$$

where λ_i, $i = 1, 2, \ldots, 6$ are predetermined levels. These constraints are called *modality constraints.* Here it may be emphasized that in this approach, the solution does not satisfy the constraints perfectly but rather it satisfies only up to a predetermined level.

Now similar to the chance constrained programming (Stancu-Minasian[70]), the modality constraints are also reduced to their deterministic equivalents. This we describe for the case where $F(x) = \sum A_j x_j$, $G(x) = \sum B_j x_j$, ζ is given by "$\tilde{\geq}$" and A_j and B_j, $(j = 1, 2, \ldots, n)$, are *L-R* fuzzy numbers defined by

$$\mu_{A_j}(x) = \begin{cases} L_{A_j}\left(\dfrac{a_j^1 - x}{\alpha_j^1}\right), x \le a_j^1, \\ 1 \qquad\qquad\quad , a_j^1 \le x \le a_j^2, \\ R_{A_j}\left(\dfrac{x - a_j^2}{\alpha_j^2}\right), x \ge a_j^2, \end{cases}$$

and

$$\mu_{B_j}(x) = \begin{cases} L_{B_j}\left(\dfrac{b_j^1 - x}{\beta_j^1}\right), x \le b_j^1, \\ 1 \qquad\qquad\quad , b_j^1 \le x \le b_j^2, \\ R_{B_j}\left(\dfrac{x - b_j^2}{\beta_j^2}\right), x \ge b_j^2, \end{cases}$$

respectively.

Here L_{A_j}, R_{A_j}, L_{B_j} and R_{B_j} are reference functions. Let us recall that a reference function $L : [0, \infty) \to [0, 1]$ is a strictly decreasing and upper semi-continuous function such that $L(0) = 1$ and $\lim_{x\to\infty} L(x) = 1$. Thus L-R fuzzy numbers A_j and B_j are denoted by $(a_j^1, a_j^2, \alpha_j^1, \alpha_j^2)$ and $(b_j^1, b_j^2, \beta_j^1, \beta_j^2)$ respectively.

Then the preference constraint $F(x)\zeta G(x)$ has six modality constraints representations whose deterministic equivalents are

$$\mu_{P_f P_l \zeta_i}(F_i(x), G_i(x)) \ge \lambda \Leftrightarrow \sum_{j=1}^{n} a_j^R(\lambda)x_j \ge \sum_{j=1}^{n} b_j^L(\lambda)x_j + \nu^*(\lambda)$$

$$\left.\begin{array}{l} \mu_{P_f N_l \zeta_i}(F_i(x), G_i(x)) \ge \lambda \\ \mu_{N_l P_f \zeta_i}(F_i(x), G_i(x)) \ge \lambda \end{array}\right\} \Leftrightarrow \sum_{j=1}^{n} a_j^R(\lambda)x_j \ge \sum_{j=1}^{n} b_j^R(1-\lambda)x_j + \nu^*(\lambda)$$

$$\left.\begin{array}{l} \mu_{N_f P_l \zeta_i}(F_i(x), G_i(x)) \ge \lambda \\ \mu_{P_l N_f \zeta_i}(F_i(x), G_i(x)) \ge \lambda \end{array}\right\} \Leftrightarrow \sum_{j=1}^{n} a_j^L(1-\lambda)x_j \ge \sum_{j=1}^{n} b_j^L(\lambda)x_j + \nu^*(\lambda)$$

$$\mu_{N_f N_l \zeta_i}(F_i(x), G_i(x)) \ge \lambda \Leftrightarrow \sum_{j=1}^{n} a_j^L(1-\lambda)x_j \ge \sum_{j=1}^{n} b_j^R(1-\lambda)x_j + \nu^*(\lambda)$$

where a_j^L, a_j^R, b_j^L b_j^R and ν^* are defined by

$$
\begin{aligned}
a_j^L(\lambda) &= a_j^1 - \alpha_j^1 L_{A_j}^*(\lambda), & j = 1,2,\ldots,n,\\
a_j^R(\lambda) &= a_j^2 + \alpha_j^2 R_{A_j}^*(\lambda), & j = 1,2,\ldots,n,\\
b_j^L(\lambda) &= b_j^1 - \beta_j^1 L_{B_j}^*(\lambda), & j = 1,2,\ldots,n,\\
b_j^R(\lambda) &= b_j^2 + \beta_j^1 R_{B_j}^*(\lambda), & j = 1,2,\ldots,n,\\
L_{A_j}^*(\lambda) &= \sup\left\{x \mid L_{A_j}(x) \geq \lambda\right\}, & j = 1,2,\ldots,n,\\
R_{A_j}^*(\lambda) &= \sup\left\{x \mid R_{A_j}(x) \geq \lambda\right\}, & j = 1,2,\ldots,n,\\
L_{B_j}^*(\lambda) &= \sup\left\{x \mid L_{B_j}(x) \geq \lambda\right\}, & j = 1,2,\ldots,n,\\
R_{B_j}^*(\lambda) &= \sup\left\{x \mid R_{B_j}(x) \geq \lambda\right\}, & j = 1,2,\ldots,n,
\end{aligned}
$$

and

$$\nu^*(\lambda) = \inf\left\{x - y \mid \mu_{\zeta_i}(x,y) = \nu(x-y) \geq \lambda\right\}.$$

Here it may be noted that for full derivation of these deterministic equivalents we have to refer to Inuiguchi et al. [28].

We next discuss the treatment of the objective function in the context of modality constrained programming. Although, in the literature, there are various approaches in this regard e.g. transformation of objective function into constraints, optimization of modalities, optimization of the fractile and minimization of ambiguity etc. Here we discuss only the one which transform the objective function into a constraint. In this context, we note that the fuzzy mathematical programming problem (*FMP*) can be expressed as a system of fuzzy constraints because the optimization of $\Phi(x)$ can be transformed as $\Phi(x)\zeta_0\Phi_0$ where Φ_0 is an appropriate fuzzy target. Therefore we can take the following formulation for (*FMP*)

(*FMP*1) Find x such that

$$
\begin{aligned}
F_i(x)\ \zeta_i\ G_i(x),\ i = 0,1,2,\ldots,m,\\
x \geq 0,
\end{aligned}
$$

where $F_0(x) = \Phi(x)$ and $G_0(x) = \Phi_0$.

In (*FMP*1) we have to find a solution x for which the degree of satisfaction of all constraints is the most. This leads to the following modality constrained programming problem (*MCP*)

(*MCP*) max $\Psi(\lambda_{1,1},\ldots,\lambda_{1,m},\ldots,\lambda_{6,1},\ldots,\lambda_{6,m})$

subject to,

$$\begin{aligned}
\mu_{P_fP_l\zeta_i}(F_i(x),G_i(x)) &\geq \lambda_{1,i},\ (i=1,2,\ldots,m),\\
\mu_{P_fN_l\zeta_i}(F_i(x),G_i(x)) &\geq \lambda_{2,i},\ (i=1,2,\ldots,m),\\
\mu_{N_fP_l\zeta_i}(F_i(x),G_i(x)) &\geq \lambda_{3,i},\ (i=1,2,\ldots,m),\\
\mu_{N_fN_l\zeta_i}(F_i(x),G_i(x)) &\geq \lambda_{4,i},\ (i=1,2,\ldots,m),\\
\mu_{N_lP_f\zeta_i}(F_i(x),G_i(x)) &\geq \lambda_{5,i},\ (i=1,2,\ldots,m),\\
\mu_{P_lN_f\zeta_i}(F_i(x),G_i(x)) &\geq \lambda_{6,i},\ (i=1,2,\ldots,m),\\
x &\geq 0,
\end{aligned}$$

where Ψ is a nondecreasing function representing the degree of satisfaction. In practice "$\Psi(\lambda_{1,1},\ldots,\lambda_{1,m},\ldots,\lambda_{6,1},\ldots,\lambda_{6,m})$" is taken as "min $(\lambda_{1,1},\ldots,\lambda_{1,m},\ldots,\lambda_{6,1},\ldots,\lambda_{6,m})$"

and in that case (MCP) reduces to

$(MCP1)$ max λ

subject to,

$$\begin{aligned}
\mu_{P_fP_l\zeta_i}(F_i(x),G_i(x)) &\geq \lambda_{1,i},\ (i=1,2,\ldots,m),\\
\mu_{P_fN_l\zeta_i}(F_i(x),G_i(x)) &\geq \lambda_{2,i},\ (i=1,2,\ldots,m),\\
\mu_{N_fP_l\zeta_i}(F_i(x),G_i(x)) &\geq \lambda_{3,i},\ (i=1,2,\ldots,m),\\
\mu_{N_fN_l\zeta_i}(F_i(x),G_i(x)) &\geq \lambda_{4,i},\ (i=1,2,\ldots,m),\\
\mu_{N_lP_f\zeta_i}(F_i(x),G_i(x)) &\geq \lambda_{5,i},\ (i=1,2,\ldots,m),\\
\mu_{P_lN_f\zeta_i}(F_i(x),G_i(x)) &\geq \lambda_{6,i},\ (i=1,2,\ldots,m),\\
x &\geq 0,
\end{aligned}$$

In the special case, when A_j and B_j are triangular fuzzy numbers i.e. $A_j=(a_j,\alpha_j^1,\alpha_j^2)$ and $B_j=(b_j,\beta_j^1,\beta_j^2)$ with

$$\mu_{A_j}(x)=\begin{cases}\dfrac{x-a_j+\alpha_j^1}{\alpha_j^1}, & a_j-\alpha_j^1\leq x\leq a_j,\\[2mm] \dfrac{a_j+\alpha_j^2-x}{\alpha_j^2}, & a_j\leq x\leq a_j+\alpha_j^2,\\[2mm] 0, & \text{otherwise,}\end{cases}$$

and

$$\mu_{B_j}(x)=\begin{cases}\dfrac{x-b_j+\beta_j^1}{\beta_j^1}, & b_j-\beta_j^1\leq x\leq b_j,\\[2mm] \dfrac{b_j+\beta_j^2-x}{\beta_j^2}, & b_j\leq x\leq b_j+\beta_j^2,\\[2mm] 0, & \text{otherwise,}\end{cases}$$

we have,

$$a_j^L(\lambda) = a_j - \alpha_j^1(1-\lambda),$$
$$a_j^R(\lambda) = a_j + \alpha_j^2(1-\lambda),$$
$$b_j^L(\lambda) = b_j - \beta_j^1(1-\lambda),$$
$$b_j^R(\lambda) = b_j + \beta_j^2(1-\lambda).$$

Hence for the fuzzy preference relation '$\gtrsim$' the deterministic equivalents of modality constraints will be somewhat simplified. This we illustrate with the help of following example.

Example 10.4.1. (Inuiguichi et al. [28]). Let us consider the following fuzzy linear programming problem:

$$\begin{aligned} &\tilde{\max} \quad C_1x_2 + C_2x_2 \\ &\text{subject to,} \\ &\quad A_1 \gtrsim B_{i1}x_1 + B_{i2}x_2, \ (i = 1,2,3), \\ &\quad x_1, x_2 \geq 0, \end{aligned}$$

where C_i's, A_i's and B_{ij}'s are triangular fuzzy numbers i.e. $C_1 = (4, 0.5, 0.5)$, $C_2 = (5, 1, 1)$, $A_1 = (330, 10, 10)$, $A_2 = (440, 5, 5)$, $A_3 = (230, 2, 2)$, $B_{11} = (2.5, 0.7, 0.7)$, $B_{12} = (5, 0.5, 0.5)$, $B_{21} = (5, 0.9, 0.9)$, $B_{22} = (6, 1.2, 1.2)$, $B_{31} = (3, 0.6, 0.6)$ and $B_{11} = (2, 0.4, 0.4)$. Further the fuzzy preference relation $\gtrsim_{p_i}$ is defined as follows:

$$\mu_{\gtrsim_{p_i}}(x,y) = \begin{cases} 1, & x \geq y, \\ 1 + \dfrac{x-y}{p_i}, & x \geq y - p_i, \\ 0, & \text{otherwise}, \end{cases}$$

where $p_1 = 10$, $p_2 = 15$ and $p_3 = 8$.
In order to transform the objective function into a constraint, we use the given target value $\Phi_0 = 400$ along with the tolerance $p_0 = 30$ and employ the membership function of the fuzzy relation $\gtrsim_{p_0}$ as described above. This reduces the given fuzzy optimization problem to the following system of fuzzy constraints

$$\begin{aligned} C_1x_2 + C_2x_2 &\gtrsim_{p_0} 400, \\ A_i &\gtrsim_{p_i} B_{i1}x_1 + B_{i2}x_2, \ (i = 1,2,3), \\ x_1, x_2 &\geq 0. \end{aligned}$$

Now this problem will be formulated by the following modality constrained programming problem:

$$\begin{aligned} &\max \quad \lambda \\ &\text{subject to,} \\ &\mu_{P_f P_l \gtrsim_0}(C_1 x_1 + C_2 x_2, 400) \geq \lambda, \\ &\mu_{P_f N_l \gtrsim_1}(A_1, B_{11}x_1 + B_{12}x_2) \geq \lambda, \\ &\mu_{N_f P_l \gtrsim_2}(A_2, B_{21}x_1 + B_{22}x_2) \geq \lambda, \\ &\mu_{N_f N_l \gtrsim_3}(A_3, B_{31}x_1 + B_{32}x_2) \geq \lambda, \\ &x_1, x_2 \geq 0, \end{aligned}$$

whose deterministic equivalent is

$$\begin{aligned} &\max \quad \lambda \\ &\text{subject to,} \\ &(4.5 - 0.5\lambda)x_1 + (6 - \lambda)x_2 \geq 370 + 30\lambda, \\ &(2.5 + 0.7\lambda)x_1 + (5 + 0.5\lambda)x_2 \geq 350 - 20\lambda, \\ &(4.1 - 0.9\lambda)x_1 + (4.8 + 1.2\lambda)x_2 \geq 455 - 20\lambda, \\ &(3 + 0.6\lambda)x_1 + (2 + 0.4\lambda)x_2 \geq 238 - 10\lambda, \\ &x_1, x_2 \geq 0. \end{aligned}$$

The above problem can be solved by a suitable numerical (crisp) optimization technique to get $x_1^* = 42.99$, $x_2^* = 41.42$ and $\lambda^* = 0.77$.

10.5 Valued relations and their fuzzy extensions

In this section we present a fuzzy relation approach for studying fuzzy linear programming problems. This approach is due to Inuiguchi et al. [29] and is quite general so as to include many well known models available in the literature.

Let X and Y be arbitrary non-empty sets. Then we know that a binary relation $\mathcal{P}$ between the elements of X and Y is a subset of the cartesian product $X \times Y$ i.e. $\mathcal{P} \subset X \times Y$. In the following we study various extensions of this definition to fuzzy scenario. For this let $F(X)$, $F(Y)$ and $F(X \times Y)$ be set of all fuzzy subsets of X, Y and $X \times Y$ respectively.

Definition 10.5.1 (Valued relation). *A fuzzy subset $\mathcal{P} \subset F(X \times Y)$ is called a valued relation on $X \times Y$ i.e. $\mathcal{P} : X \times Y \to [0,1]$.*

Definition 10.5.2 (Fuzzy relation). *A valued relation $\mathcal{P}$ on $F(X) \times F(Y)$ is called a fuzzy relation on $X \times Y$ and it is denoted by $\tilde{\mathcal{P}}$. In other words, a fuzzy reaction $\tilde{\mathcal{P}}$ on $X \times Y$ is a fuzzy subset of $F(X) \times F(Y)$ i.e. $\tilde{\mathcal{P}} : F(X) \times F(Y) \to [0,1]$.*

Definition 10.5.3 (Fuzzy extension of a valued relation $\mathcal{P}$). *Let $\mathcal{P}$ be a valued relation on $X \times Y$. A fuzzy relation $\tilde{Q}$ on $X \times Y$ with $\mu_{\tilde{Q}}(x, y) = \mu_{\mathcal{P}}(x, y)\ \forall x \in X,\ y \in Y$ is called a fuzzy extension of the relation $\mathcal{P}$.*

Here, it may be recalled that x and y in $\mu_{\mathcal{P}}(x, y)$ belong to the sets X and Y and, x and y in $\mu_{\tilde{Q}}(x, y)$ are elements of $F(X)$ and $F(Y)$ respectively with the understanding that $x \in X$ (respectively $y \in Y$) can be considered a fuzzy subset of X (respectively Y) with the characteristic function χ_x (respectively χ_y) as the membership function.

Definition 10.5.4 (Dual fuzzy extension of a valued relation $\mathcal{P}$). *Let X be a non-empty set and $\mathcal{P}$ be a valued relation on X. Let $\mu_{c_{\mathcal{P}}}$ be the membership function of the valued relation $c_{\mathcal{P}}$ given by*

$$\mu_{c_{\mathcal{P}}}(x, y) = 1 - \mu_{\mathcal{P}}(x, y) \text{ for all } x,\ y \in X.$$

Let $\tilde{Q}$ be a fuzzy extension of relation $c_{\mathcal{P}}$. Then a fuzzy relation $\tilde{Q}^d$ on X defined by

$$\mu_{\tilde{Q}^d}(A, B) = 1 - \mu_{\tilde{Q}}(B, A) \text{ for all } A, B \in F(X),$$

is called the dual fuzzy extension of the valued relation $\mathcal{P}$.

For a given valued relation $\mathcal{P}$, there might be considered many fuzzy extensions and hence many dual extensions. For an specific valued relation $\mathcal{P}$, its fuzzy extension is denoted by $\tilde{\mathcal{P}}$; and its dual fuzzy extension is denoted by $\tilde{Q}^d$.

Let X, Y be non-empty sets, T be a t-norm and S be a t-conorm. Let $\mathcal{P}$ be a valued relation on $X \times Y$. This valued relation $\mathcal{P}$ is extended to the following fuzzy extensions by t-norms and t-conorms. These extension are inspired by Dubois and Prade [16] as given in Section 10.3.

(i) A fuzzy relation $\tilde{\mathcal{P}}^T$ on $X \times Y$ defined for all $A \in F(X)$ and $B \in F(Y)$ by

$$\mu_{\tilde{\mathcal{P}}^T}(A, B) = \sup_{x \in X, y \in Y} \{T(\mu_{\mathcal{P}}(x, y), T(\mu_A(x), \mu_B(y))\}$$

(ii) A fuzzy relation $\tilde{\mathcal{P}}_S$ on $X \times Y$ defined for all $A \in F(X)$ and $B \in F(Y)$ by

$$\mu_{\tilde{\mathcal{P}}_S}(A, B) = \inf_{x \in X, y \in Y} \left\{S\big(S\big(1 - \mu_A(x), 1 - \mu_B(y)\big), \mu_{\mathcal{P}}(x, y)\big)\right\}$$

(iii) A fuzzy relation $\tilde{\mathcal{P}}^{T,S}$ on $X\times Y$ defined for all $A \in F(X)$ and $B \in F(Y)$ by

$$\mu_{\tilde{\mathcal{P}}^{T,S}}(A,B) = \sup_{x\in X}\left\{\inf_{y\in Y}\left\{T(\mu_A(x), S(1-\mu_B(y), \mu_{\mathcal{P}}(x,y)))\right\}\right\}$$

(iv) A fuzzy relation $\tilde{\mathcal{P}}_{T,S}$ on $X\times Y$ defined for all $A \in F(X)$ and $B \in F(Y)$ by

$$\mu_{\tilde{\mathcal{P}}_{T,S}}(A,B) = \inf_{y\in Y}\left\{\sup_{x\in X}\left\{S(T(\mu_A(x), \mu_{\mathcal{P}}(x,y))), 1-\mu_B(x,y))\right\}\right\}$$

(v) A fuzzy relation $\tilde{\mathcal{P}}^{S,T}$ on $X\times Y$ defined for all $A \in F(X)$ and $B \in F(Y)$ by

$$\mu_{\tilde{\mathcal{P}}^{S,T}}(A,B) = \sup_{y\in Y}\left\{\inf_{x\in X}\left\{T(S(1-\mu_A(x), \mu_{\mathcal{P}}(x,y)), \mu_B(y))\right\}\right\}$$

(vi) A fuzzy relation $\tilde{\mathcal{P}}_{S,T}$ on $X\times Y$ defined for all $A \in F(X)$ and $B \in F(Y)$ by

$$\mu_{\tilde{\mathcal{P}}_{S,T}}(A,B) = \inf_{x\in X}\left\{\sup_{y\in Y}\left\{S(1-\mu_A(x), T(\mu_B(x,y), \mu_{\mathcal{P}}(x,y)))\right\}\right\}$$

All the above fuzzy extensions of a valued relation are fuzzy extensions in the sense of Definition 10.5.3. In the above, the fuzzy relations $\tilde{\mathcal{P}}^T$ and $\tilde{\mathcal{P}}_S$ are respectively called the T-fuzzy extension and S-fuzzy extension of relation $\mathcal{P}$.

Let $\mathcal{P}$ be the classical binary relation "less than or equal to" denoted by '$\leq$' on $\mathbb{R}$, T be the 'min' and S be the 'max'. Let us denote $\tilde{\mathcal{P}}^T$ by $\tilde{\leqq}^M$ and $\tilde{\mathcal{P}}_S$ by $\tilde{\leqq}_M$. Then

$$\mu_{\tilde{\leqq}^M}(A,B) = \sup\left\{\min\left(\mu_A(x), \mu_B(y), \mu_{\leq}(x,y)\right) : x \in R, y \in R\right\},$$

$$\mu_{\tilde{\leqq}_M}(A,B) = \inf\left\{\max\left(1-\mu_A(x), 1-\mu_B(y), \mu_{\leq}(x,y)\right) : x \in R, y \in R\right\}.$$

Here we note that $\tilde{\leqq}^M$ and $\tilde{\leqq}_M$ are fuzzy relations which are dual to each other.

We now state below given theorem to be used in the next section. For proof of this theorem we shall refer to Inuiguchi et al. [29]. Let us recall that a fuzzy set in $\mathbb{R}$ is called *compact* if all its α-cuts are *compact* for all $\alpha \in (0,1]$.

Theorem 10.5.1 *Let A, $B \in F(\mathbb{R})$ be normal and compact, $T = \min$, $S = \max$ and $\alpha \in (0,1)$. Then*

(i) $\mu_{\tilde{\leqq}^T}(A,B) \geq \alpha \Leftrightarrow \inf[A]_\alpha \leq \sup[B]_\alpha$,
(ii) $\mu_{\tilde{\leqq}_S}(A,B) \geq \alpha \Leftrightarrow \sup(A)_{1-\alpha} \leq \inf(B)_{1-\alpha}$,
(iii) $\mu_{\tilde{\leqq}^{T,S}}(A,B) \geq \alpha \Leftrightarrow \mu_{\tilde{\leqq}_{T,S}}(A,B) \geq \alpha \Leftrightarrow \sup(A)_{1-\alpha} \leq \sup[B]_\alpha$,
(iv) $\mu_{\tilde{\leqq}^{S,T}}(A,B) \geq \alpha \Leftrightarrow \mu_{\tilde{\leqq}_{S,T}}(A,B) \geq \alpha \Leftrightarrow \inf[A]_\alpha \leq \inf(B)_{1-\alpha}$.

10.6 Fuzzy linear programming via fuzzy relations

Before introducing the fuzzy linear programming problem to be studied in this section, we introduce the following definitions.

Definition 10.6.1 (Quasiconcave function). *Let* $X \subset \mathbb{R}$ *and* $f : \mathbb{R} \to [0,1]$. *Then* f *is called quasiconcave on* X *if*

$$f(\lambda x + (1-\lambda)y) \geq \min(f(x), f(y)),$$

for all $x, y \in X$ *and for all* $\lambda \in (0,1)$ *with* $\lambda x + (1-\lambda)y \in X$.

Definition 10.6.2 (Strictly quasiconcave function). *Let* $X \subset \mathbb{R}$ *and* $f : \mathbb{R} \to [0,1]$. *Then* f *is called strictly quasiconcave on* X *if*

$$f(\lambda x + (1-\lambda)y) > \min(f(x), f(y)),$$

for all $x, y \in X$, $x \neq y$ *and every* $\lambda \in (0,1)$ *with* $\lambda x + (1-\lambda)y \in X$.

Definition 10.6.3 (Semistrictly quasiconcave function). *Let* $X \subset \mathbb{R}$ *and* $f : \mathbb{R} \to [0,1]$. *Then* f *is called semistrictly quasiconcave on* X *if* f *is quasiconcave on* X *and*

$$f(\lambda x + (1-\lambda)y) > \min(f(x), f(y)),$$

for all $x, y \in X$, *and every* $\lambda \in (0,1)$ *with* $\lambda x + (1-\lambda)y \in X$, $f(\lambda x + (1-\lambda)y) > 0$ *and* $f(x) \neq f(y)$.

Definition 10.6.4 (Fuzzy quantity). *Let* S *be a fuzzy subset of* $\mathbb{R}$. *Then* S *is called a fuzzy quantity if* S *is normal, compact and has semi strictly quasiconcave membership function.*

In this context we may note that the characteristic functions of crisp subsets of $\mathbb{R}$ are semistrictly qausiconcave on $\mathbb{R}$ but not strictly qausiconcave on $\mathbb{R}$. Let $F_0(\mathbb{R})$ denote the set of all fuzzy quantities on $\mathbb{R}$. Then it may be checked that $F_0(\mathbb{R})$ includes all (crisp) real numbers, crisp intervals and triangular fuzzy numbers etc. but not every fuzzy number is a fuzzy quantity.

Lemma 10.6.1. *Let* $\tilde{a}_j$ *be a fuzzy quantity i.e.* $\tilde{a}_j \in F_0(\mathbb{R})$ *and* $x_j \geq 0$, $(j = 1, 2, \ldots, n)$. *Then* $\sum_{j=1}^{n} \tilde{a}_j x_j$ *is also a fuzzy quantity. Here* $\sum_{j=1}^{n} \tilde{a}_j x_j$ *is understood to be a fuzzy set on* $\mathbb{R}$ *whose membership function is defined as per Zadeh extension principle.*

We now consider the fuzzy linear programming problem

$$(FLPP) \qquad \tilde{\max} \quad \sum_{j=1}^{n} \tilde{c}_j x_j$$

subject to,

$$\sum_{j=1}^{n} \tilde{a}_{ij} x_j \; \tilde{P}_i \; \tilde{b}_i, \; (i = 1, 2, \ldots, m),$$

$$x_j \geq 0, \quad (j = 1, 2, \ldots, n),$$

where $\tilde{c}_j$, $\tilde{a}_{ij}$ and $\tilde{b}_i$ are fuzzy quantities, whose membership functions are $\mu_{\tilde{c}_j} : R \to [0,1]$, $\mu_{\tilde{a}_{ij}} : R \to [0,1]$ and $\mu_{\tilde{b}_i} : R \to [0,1]$ respectively. Here, $\tilde{P}_i$ is a fuzzy relation on R which is used to "compare" the fuzzy quantities $\sum_{j=1}^{n} \tilde{a}_{ij} x_j$ and $\tilde{b}_i$ $(i = 1, 2, \ldots, m)$. Also we have to understand the meaning of $\tilde{\max}$ of a fuzzy quantity and that we shall discuss later in this sequel.

For a given x, the objective function value $\sum_{j=1}^{n} \tilde{c}_j x_j$ and the left hand side values of the constraints $\sum_{j=1}^{n} \tilde{a}_{ij} x_j$ are fuzzy quantities whose respective membership functions are defined by using Zadeh's extension principle as follows:

$$\mu_{\tilde{f}}(t) = \begin{cases} \sup\limits_{c_1, \ldots, c_n \in R, \; \sum_{j=1}^{n} c_j x_j = t} \Big\{ T\Big(\mu_{\tilde{c}_1}(c_1), \ldots, \mu_{\tilde{c}_n}(c_n) \Big) \Big\}, & x \neq 0 \; or \; t = 0 \\ 0, & \text{otherwise} \, , \end{cases}$$

and

$$\mu_{\tilde{g}_i}(t) = \begin{cases} \sup\limits_{a_1, \ldots a_n, \in R, \; \sum_{j=1}^{n} a_{ij} x_j = t} \Big\{ T\Big(\mu_{\tilde{a}_{i1}}(a_1), \ldots, \mu_{\tilde{a}_{in}}(a_n) \Big) \Big\}, & x \neq 0 \; or \; t = 0 \\ 0, & \text{otherwise.} \end{cases}$$

Next in this sequel is to understand the meaning of the feasible solution to the fuzzy linear programming problem $(FLPP)$.

Definition 10.6.5 (Feasible solution). *Let* $\mu_{\tilde{a}_{ij}}, \mu_{\tilde{b}_i} : \mathbb{R} \longrightarrow [0,1]$, $(i = 1,2,\ldots,m)$, $(j = 1,2,\ldots,n)$ *be membership functions of fuzzy quantities* $\tilde{a}_{ij}$ *and* $\tilde{b}_i$ *respectively. Let* $\tilde{P}_i$, $(i = 1,2,\ldots,m)$ *be fuzzy relations on* $\mathbb{R}$ *and let* T^A *and* T *be t-norms. A fuzzy set* $\tilde{X}$, *with membership function given by*

$$\mu_{\tilde{X}}(x) = \begin{cases} T^A\left(\mu_{\tilde{P}_1}\Big(\sum_{j=1}^{n} \tilde{a}_{1j}x_j, \tilde{b}_1\Big), \ldots, \mu_{\tilde{P}_m}\Big(\sum_{j=1}^{n} \tilde{a}_{mj}x_j, \tilde{b}_n\Big)\right), & \\ & x_i \geq 0, 1 \leq j \leq n, \\ 0, & \text{otherwise}, \end{cases}$$

is called a feasible solution of the problem (FLPP). Here t-norm T^A *is used as an aggregation operator, while the t-norm* T *is used for extending arithmetic operations*

Definition 10.6.6 (α-Feasible solution). *Let* $\tilde{X}$ *be a feasible solution. Then for* $\alpha \in (0,1]$, *a vector* $x \in [\tilde{X}]_\alpha$ *is called the* α*-feasible solution of (FLPP).*

Definition 10.6.7 (Max-feasible solution). *Let* $\tilde{X}$ *be a feasible solution. A vector* $\bar{x} \in R^n$ *such that* $\mu_{\tilde{X}}(\bar{x}) = Hgt(\tilde{X})$ *is called a max-feasible solution.*

From the above definitions we note that the feasible solution $\tilde{X}$ of (*FLPP*) is a fuzzy set but a α-feasible solution is a vector $\bar{x} \in \mathbb{R}^n$ which belongs to α-cut of the feasible solution $\tilde{X}$. Further a max-feasible solution is a α-feasible solution for which $\alpha = \text{Hgt}(\tilde{X})$. If we agree to denote by $\tilde{X}_i$ the fuzzy set of $\mathbb{R}^n$ with the membership function $\mu_{\tilde{X}_i}$ defined by

$$\mu_{\tilde{X}_i} = \mu_{\tilde{P}_i}(\tilde{a}_{i1}x_1 + \ldots + \tilde{a}_{in}x_n, \tilde{b}_i),$$

then it is interpreted as the ith fuzzy constraint $(i = 1,2,\ldots,m)$. The feasible solution $\tilde{X}$ is obtained by aggregating all these fuzzy sets $\tilde{X}_i$ by the t-norm T^A.

Theorem 10.6.1 *Let* $\tilde{a}_{ij}$ *and* $\tilde{b}_i$ *be fuzzy quantities for* $i = (1,2,\ldots,m)$, $j = (1,2,\ldots,n)$, *and* $x_i \geq 0$. *Let* $T = \min$, $S = \max$, $\alpha \in (0,1)$ *and '$\leq$' be the usual binary relation "less than or equal to". Then for* $i = (1,2,\ldots,m)$

(i)

$$\mu_{\tilde{\geq}^T}\Big(\sum_{j=1}^{n}\tilde{a}_{ij}x_j,\tilde{b}_i\Big)\geq\alpha \ \text{ if and only if } \sum_{j=1}^{n}\bar{a}_{ij}^L(\alpha)x_j\leq\bar{b}_i^R(\alpha)$$

(ii)

$$\mu_{\tilde{\leq}_S}\Big(\sum_{j=1}^{n}\tilde{a}_{ij}x_j,\tilde{b}_i\Big)\geq\alpha \ \text{ if and only if } \sum_{j=1}^{n}a_{ij}^R(1-\alpha)x_j\leq b_i^L(1-\alpha)$$

(i)

$$\mu_{\tilde{\leq}T,S}\Big(\sum_{j=1}^{n}\tilde{a}_{ij}x_j,\tilde{b}_i\Big)\geq\alpha \ \text{ if and only if } \sum_{j=1}^{n}a_{ij}^R(1-\alpha)x_j\leq\bar{b}_i^R(\alpha)$$

where $\mu_{\tilde{\leq}T,S}=\mu_{\tilde{\leq}^{T,S}}=\mu_{\tilde{\leq}_{T,S}}$, *and*

(iv)

$$\mu_{\tilde{\leq}S,T}\Big(\sum_{j=1}^{n}\tilde{a}_{ij}x_j,\tilde{b}_i\Big)\geq\alpha \ \text{ if and only if } \sum_{j=1}^{n}a_{ij}^L(\alpha)x_j\leq\bar{b}_i^L(1-\alpha)$$

where $\mu_{\tilde{\leq}S,T}=\mu_{\tilde{\leq}^{S,T}}=\mu_{\tilde{\leq}_{S,T}}$.

Here for given $\alpha\in[0,1]$, $(i=1,2,\ldots,m)$, $(j=1,2,\ldots,n)$ *we have*

$$\bar{a}_{ij}^L(\alpha)=\inf\Big\{a\in\mathbb{R}:a\in\big[\tilde{a}_{ij}\big]_\alpha\Big\},$$

$$a^L{}_{ij}(\alpha)=\inf\Big\{a\in\mathbb{R}:a\in\big(\tilde{a}_{ij}\big)_\alpha\Big\},$$

$$\bar{a}_{ij}^R(\alpha)=\sup\Big\{a\in\mathbb{R}:a\in\big[\tilde{a}_{ij}\big]_\alpha\Big\},$$

$$a^R{}_{ij}(\alpha)=\sup\Big\{a\in\mathbb{R}:a\in\big(\tilde{a}_{ij}\big)_\alpha\Big\}.$$

similarly $\bar{b}_i^L(\alpha)$, $b^L{}_i(\alpha)$, $\bar{b}_i^R(\alpha)$ *and* $b^R{}_i(\alpha)$ *are defined.*

We shall not prove the above theorem here, but for the same one can refer Inuiguchi et al. ([26], [27], [28], [29]).

For a given x, the objective function value $f(x)=\sum\tilde{c}_jx_j$ is a fuzzy quantity. In order to "maximize" the fuzzy objective function we need a suitable ordering on $F_0(\mathbb{R})$ and also a suitable exogenously chosen fuzzy goal to which the fuzzy values of the objective function can be compared by an associated fuzzy relation $\tilde{P}_0$ on $\mathbb{R}$. Consequently, if $\tilde{d}\in F_0(\mathbb{R})$ is the chosen fuzzy goal then the fuzzy objective $\tilde{c}_1x_1 + \ldots + \tilde{c}_nx_n$ is treated as another constraint

$$\tilde{c}_1x_1 + \ldots + \tilde{c}_nx_n \; \tilde{P}_0 \; \tilde{d}.$$

Therefore given (*FLPP*) can be represented as
(*FLPP*1) Find $x \in \mathbb{R}^n$ such that

$$\begin{aligned} \tilde{c}_1x_1 + \ldots + \tilde{c}_nx_n \; \tilde{P}_0 \; \tilde{d}, \\ \tilde{a}_{i1}x_1 + \ldots + \tilde{a}_{in}x_n \; \tilde{P}_i \; \tilde{b}_i, \; (i = 1, 2, \ldots, m), \\ x_j \geq 0, \;\; (j = 1, 2, \ldots, n). \end{aligned}$$

Now by utilizing the definition of feasible solution to (*FLPP*), we introduce a satisficing solution.

Definition 10.6.8 (Satisficing solution). *Let* $\mu_{\tilde{c}_j} : \mathbb{R} \to [0,1]$, $\mu_{\tilde{a}_{ij}} : \mathbb{R} \to [0,1]$ *and* $\mu_{\tilde{b}_i} : \mathbb{R} \to [0,1]$, $i = (1, 2, \ldots, m)$ $j = (1, 2, \ldots, n)$, *be membership functions of* $\tilde{c}_j$, $\tilde{a}_{ij}$ *and* $\tilde{b}_i$, *respectively. Also let* $\tilde{d} \in F_0(\mathbb{R})$ *be a fuzzy goal and* T^G *be a t-norm. Then a fuzzy set* $\tilde{X}^*$ *with the membership function* $\mu_{\tilde{X}^*}$ *defined by*

$$\mu_{\tilde{X}^*}(x) = T^G(\mu_{\tilde{P}_0}(\tilde{c}_1x_1 + \ldots + \tilde{c}_nx_n, \tilde{d}), \mu_{\tilde{X}}(x)), \; x \in \mathbb{R}^n,$$

where $\mu_{\tilde{X}}(x)$ *is the membership function of the feasible solution* $\tilde{X}$, *is called a satisficing solution.*
In the above definition the t-norm T^G *has been used for aggregation of the fuzzy set of the feasible solution* $\tilde{X}$ *with the fuzzy set of the objective* $\tilde{X}_0$ *where*

$$\mu_{\tilde{X}_0}(x) = \mu_{\tilde{P}_0}(\tilde{c}_1x_1 + \ldots + \tilde{c}_nx_n, \tilde{d}), \; x \in \mathbb{R}^n.$$

Hence, the membership function of satisficing solution $\tilde{X}^*$ *for all* $x \in \mathbb{R}^n$ *can be obtained as*

$$\mu_{\tilde{X}^*}(x) = T^G(\mu_{\tilde{X}_0}(x), \mu_{\tilde{X}}(x)).$$

For $\alpha \in (0,1]$ a vector $x \in [\tilde{X}^*]_\alpha$ is called the α-satisficing solution of fuzzy linear programming problem (*FLPP*).

A vector $x^* \in \mathbb{R}^n$ with the property

$$\mu_{\tilde{X}^*}(x^*) = Hgt(\tilde{X}^*)$$

is called the max-satificing solution.

Now, consider membership functions of the fuzzy objective and fuzzy constraints of fuzzy linear programming problem (*FLPP*) as

$$\mu_{\tilde{X}_0}(x) = \mu_{\tilde{P}_0}(\tilde{c}_1 x_1 + \ldots + \tilde{c}_n x_n, \tilde{d}),$$

and

$$\mu_{\tilde{X}_i}(x) = \mu_{\tilde{P}_i}(\tilde{a}_{i1} x_1 + \ldots + \tilde{a}_{in} x_n, \tilde{b}_i),$$

for each $x \in \mathbb{R}^n$ and $i = (1, 2, \ldots, m)$. Also, $T^G = T^A = \min$. Then the following theorem can be stated (Inuiguchi et al. [29]).

Theorem 10.6.2 *A vector* $(\lambda^*, x^*) \in \mathbb{R}^{n+1}$ *is an optimal solution of the optimization problem*

$$\begin{array}{ll} \max & \lambda \\ \text{subject to,} & \\ & \lambda \leq \mu_{\tilde{X}_i}(x), \ (i = 0, 1, \ldots, m), \\ & x_j \geq 0, \qquad (j = 1, \ldots, n), \end{array}$$

if and only if $x^* \in \mathbb{R}^n$ *is a max-satisficing solution of fuzzy linear programming problem* (*FLPP*).

10.7 Duality in fuzzy linear programming via fuzzy relations

This section is devoted to the study of duality in fuzzy linear programming by using fuzzy (valued) relations. Here we introduce a pair of primal dual fuzzy linear programming problems (*FLPP*) and (*DFLPP*) as given below

(*FLPP*)

$$\begin{array}{ll} \widetilde{\max} & \tilde{c}_1 x_1 + \ldots + \tilde{c}_n x_n \\ \text{subject to,} & \\ & \tilde{a}_{i1} x_1 + \ldots + \tilde{a}_{in} x_n \ \tilde{\mathcal{P}}_i \ \tilde{b}_i, \quad (i = 1, 2, \ldots, m), \\ & x_j \geq 0, \quad (j = 1, 2, \ldots, n), \end{array}$$

and

(*DFLPP*)

$$\begin{array}{ll} \widetilde{\min} & \tilde{b}_1 y_1 + \ldots + \tilde{b}_n y_n \\ \text{subject to,} & \\ & \tilde{a}_{i1} y_1 + \ldots + \tilde{a}_{in} y_n \ \tilde{Q}_i^d \ \tilde{c}_i, \quad (j = 1, 2, \ldots, n), \\ & y_j \geq 0, \quad (i = 1, 2, \ldots, m). \end{array}$$

Here the coefficients $\tilde{c}_j$, $\tilde{a}_{ij}$ and $\tilde{b}_i$ are as introduced in Section 10.5. The above (*FLPP*) is called primal fuzzy linear programming problem, while the problem (*DFLPP*) is called its dual. Also $\tilde{\mathcal{P}}_i$ is the fuzzy extension of a valued relation $\mathcal{P}_i$ on $\mathbb{R}$ and $\tilde{Q}_i^{\ d}$ is the dual fuzzy extension of $\mathcal{P}_i$.

If we consider fuzzy relation $\tilde{\mathcal{P}}$ as $\tilde{\leq}^T$ then the corresponding dual fuzzy relation $\tilde{Q}^d$ will be $\tilde{\geq}^S$, which will give us the following pair of primal dual fuzzy linear programming problems

(*FLPP*1) $\quad \tilde{\max} \quad \tilde{c}_1x_1+\ldots+\tilde{c}_nx_n$

subject to,

$$\tilde{a}_{i1}x_1+\ldots+\tilde{a}_{in}x_n \tilde{\leq}^T \tilde{b}_i, (i=1,2,\ldots,m),$$
$$x_j \geq 0, \quad (j=1,2,\ldots,n),$$

and

(*DFLPP*1) $\quad \tilde{\min} \quad \tilde{b}_1y_1+\ldots+\tilde{b}_my_m$

subject to,

$$\tilde{a}_{1j}y_1+\ldots+\tilde{a}_{mj}y_m \tilde{\geq}_S \tilde{c}_j, (j=1,2,\ldots,n),$$
$$y_j \geq 0, \quad (i=1,2,\ldots,m).$$

Theorem 10.7.1 *Let for all* $i=1,2,\ldots,m$, $j=1,2,\ldots,n$, $\tilde{c}_j$, $\tilde{a}_{ij}$ *and* $\tilde{b}_i$ *be fuzzy quantities. let* $\tilde{\leq}^T$ *be the T-fuzzy extension of the binary relation* $\leq$ *on R and* $\tilde{\geq}_S$ *be the S-fuzzy extension of the relation* $\geq$ *on R. Let* $\tilde{X}$ *be a feasible solution of* (*FLPP*1), $\tilde{Y}$ *be a feasible solution of* (*DFLPP*1), *and* $\alpha \in [0.5,1)$. *If* $x=(x_1,\ldots,x_n)\geq 0$ *belong to* $\left[\tilde{X}\right]_\alpha$ *and* $y=(y_1,\ldots,y_m)\geq 0$ *belongs to* $\left[\tilde{Y}\right]_\alpha$, *then*

$$\sum_{(j=1,2,\ldots,n)} \bar{c}_j^R(1-\alpha)x_j \leq \sum_{(i=1,2,\ldots,m)} \bar{b}_j^R(1-\alpha)y_j.$$

Proof. From the strict convexity and normality, for all $(i=1,2,\ldots,m)$, $(j=1,2,\ldots,n)$, we have $\bar{c}_j^L=c_j^L, \bar{c}_j^R=c_j^R, \bar{a}_{ij}^L=a_{ij}^L$, $\bar{a}_{ij}^R=a_{ij}^R$, $\bar{b}_i^L=b_i^L$ and $\bar{b}_i^R=b_i^R$. Let $x\in\left[\tilde{X}\right]_\alpha$ and $y\in\left[\tilde{Y}\right]_\alpha$, $x_j\geq 0$, $y_j\geq 0$ for all $(i=1,2,\ldots,m)$, $(j=1,2,\ldots,n)$. Then by Theorem 10.6.1 we obtain

$$\sum_{(i=1,2,\ldots,m)} \bar{a}_{ij}^L(1-\alpha)y_i = \sum_{(i=1,2,\ldots,m)} a_{ij}^L(1-\alpha)y_i \geq c_j^R(1-\alpha) = \sum_{(i=1,2,\ldots,m)} \bar{c}_j^R(1-\alpha).$$

Since $\alpha\geq 0.5$, it follows that $1-\alpha\leq\alpha$, hence $\left[\tilde{X}\right]_\alpha \subset \left[\tilde{X}\right]_{1-\alpha}$. Again by above Theorem 10.6.1 we obtain for all $(i=1,2,\ldots,m)$

$$\sum_{(i=1,2,\ldots,m)} \bar{a}^L_{ij}(1-\alpha)x_j \leq \bar{b}^R_i(1-\alpha).$$

Multiplying both sides of these equations by $x_j \geq 0$ and $y_j \geq 0$, respectively, and summing up the result, we obtain

$$\sum_{(j=1,2,\ldots,n)} \bar{c}^R_j(1-\alpha)x_j \leq \sum_{(j=1,2,\ldots,n)} \sum_{(i=1,2,\ldots,m)} \bar{a}^L_{ij}(1-\alpha)y_i x_j \leq \sum_{(i=1,2,\ldots,m)} \bar{b}^R(1-\alpha)y_i.$$

which is the desired result.

10.8 Duality in fuzzy LPPs with fuzzy coefficients: Wu's model

We have already discussed some models for studying duality in fuzzy linear programming problems with fuzzy coefficients e.g. the ranking function approach discussed in Chapter 6 and the fuzzy relations approach discussed here in Section 10.4. In this section we present Wu's approach [82] for studying duality in fuzzy linear programming which is based on the concept of fuzzy scalar product and looks very similar to the crisp linear programming duality. Although theoretically it looks very general, in actual practice there are certain limitations as we shall see later in this section.

Let us recall that for a fuzzy number $\tilde{a}$, its α-cut $[\tilde{a}]^\alpha$, for $\alpha \in [0,1]$, is denoted by the interval $[a^L_\alpha, a^R_\alpha]$. Therefore if $\tilde{a}$ and $\tilde{b}$ are two fuzzy numbers with α-cuts, as $[a^L_\alpha, a^R_\alpha]$, and $[b^L_\alpha, b^R_\alpha]$ respectively, then $\tilde{a}(+)\tilde{b}$ and $\tilde{a}(\cdot)\tilde{b}$ are also fuzzy numbers with respective α-cuts as

$$[\tilde{a}(+)\tilde{b}]^\alpha = [a^L_\alpha + b^L_\alpha, a^R_\alpha + b^R_\alpha],$$

and

$$[\tilde{a}(\cdot)\tilde{b}]^\alpha = [\min(a^L_\alpha b^L_\alpha, a^L_\alpha b^R_\alpha, a^R_\alpha b^L_\alpha, a^R_\alpha b^R_\alpha), \max(a^L_\alpha b^L_\alpha, a^L_\alpha b^R_\alpha, a^R_\alpha b^L_\alpha, a^R_\alpha b^R_\alpha)].$$

In the following, sometimes if there is no possibility of confusion, then the fuzzy sum $\tilde{a}(+)\tilde{b}$ and the fuzzy product $\tilde{a}(\cdot)\tilde{b}$ will be denoted by $\tilde{a} + \tilde{b}$ and $\tilde{a} \cdot \tilde{b}$ only.

Definition 10.8.1 (Nonnegative/nonpositive fuzzy number). *Let $\tilde{a}$ be a fuzzy number. Then $\tilde{a}$ is called a nonnegative fuzzy number if $\mu_{\tilde{a}}(x) = 0$ for all $x < 0$. Similarly $\tilde{a}$ is called nonpositive fuzzy number if $\mu_{\tilde{a}}(x) = 0$ for all $x > 0$.*

Here it may be observed that if $\tilde{a}$ is a nonnegative fuzzy number then a^L_α and a^R_α are nonnegative real numbers for all $\alpha \in [0,1]$. Now for the given fuzzy number $\tilde{a}$, we define a fuzzy number $\tilde{a}^+$ with the following membership function

$$\mu_{\tilde{a}^+}(r) = \begin{cases} \mu_{\tilde{a}}(r) & , r > 0, \\ 1 & , r = 0 \text{ and } \mu_{\tilde{a}}(r) < 1 \text{ for all } r > 0, \\ \mu_{\tilde{a}}(0) & , r = 0 \text{ and } \exists\, r > 0 \text{ such that } \mu_{\tilde{a}}(r) = 1, \\ 0 & , \text{otherwise.} \end{cases}$$

From the above it is clear that $\tilde{a}^+$ is a nonnegative fuzzy number. Similarly we define a nonpositive number $\tilde{a}_-$ with the following membership function

$$\mu_{\tilde{a}^-}(r) = \begin{cases} \mu_{\tilde{a}}(r) & , r < 0, \\ 1 & , r = 0 \text{ and } \mu_{\tilde{a}}(r) < 1 \text{ for all } r < 0, \\ \mu_{\tilde{a}}(0) & , r = 0 \text{ and } \exists\, r < 0 \text{ such that } \mu_{\tilde{a}}(r) = 1, \\ 0 & , \text{otherwise.} \end{cases}$$

Here it can be verified that $\tilde{a} = \tilde{a}^+(+)\tilde{a}^-$. Thus every fuzzy number $\tilde{a}$ can be written as the sum of fuzzy numbers $\tilde{a}^+$ and $\tilde{a}^-$which are respectively called the *positive part* and the *negative part* of $\tilde{a}$.

In the following we shall treat a crisp number m also as a fuzzy number $\tilde{a}$ with membership function as its characteristic function, i.e.

$$\mu_{\tilde{a}}(r) = \begin{cases} 1, & r = m, \\ 0, & r \neq m, \end{cases}$$

and use the notation $\tilde{1}(m)$ for its representation.

Definition 10.8.2 (Ordering of fuzzy numbers). *Given two fuzzy numbers $\tilde{a}$ and $\tilde{b}$ we write $\tilde{b} \geqq \tilde{a}$ if $b^L_\alpha \geq a^L_\alpha$ and $b^R_\alpha \geq a^R_\alpha$ for all $\alpha \in [0,1]$. We write $\tilde{a} \leqq \tilde{b}$ if $\tilde{b} \geqq \tilde{a}$.*

Further $\tilde{a} \geq \tilde{b}$ is defined as $\tilde{a} \geqq \tilde{b}$ and $\tilde{a} \neq \tilde{b}$, i.e. $\tilde{a} \geqq \tilde{b}$ and there exists an $\alpha \in [0,1]$ such that $a^L_\alpha > b^L_\alpha$ or $a^R_\alpha > b^R_\alpha$. Also if H is a set of fuzzy numbers then we use the symbol $H \leqq \tilde{k}$ if $\tilde{h} \leqq \tilde{k}$ for all $\tilde{h} \in H$. In a similar manner, if H and K are two sets of fuzzy numbers then $H \leqq K$ is understood as $H \leqq \tilde{k}$ for all $\tilde{k} \in K$.

We note that Definition 10.8.2 above is the same as Definition 8.2.2 except a notational change that for scalars $a, b \in \mathbb{R}$ we are writing $b \geq a$ rather than $b \geqq a$ as these are same statements. Let $F^n(\mathbb{R}) = F(\mathbb{R}) \times \ldots \times F(\mathbb{R})$ and $\tilde{x} \in F^n(\mathbb{R})$, i.e. $\tilde{x} = (\tilde{x}_1, \ldots, \tilde{x}_n)$ with $\tilde{x}_i \in F(\mathbb{R})$ for $i = 1, 2, \ldots, n$. Let $x^L_\alpha = (x^L_{1\alpha}, \ldots, x^L_{n\alpha})$, $x^R_\alpha = (x^R_{1\alpha}, \ldots, x^R_{n\alpha})$ where

$x^L_{i\alpha} = (x_i)^L_\alpha$ and $x^R_{i\alpha} = (x_i)^R_\alpha$. Then for $\tilde{x}, \tilde{y} \in F^n(\mathbb{R})$ we define $\tilde{x}(+)\tilde{y}$ as $\tilde{x}(+)\tilde{y} = (\tilde{x_1}(+)\tilde{y_1}, \ldots, \tilde{x_n}(+)\tilde{y_n})$. Also if $\tilde{x}^+ = (\tilde{x}_1^+, \ldots, \tilde{x}_n^+)$ and $\tilde{x}^- = (\tilde{x}_1^-, \ldots, \tilde{x}_n^-)$ then $\tilde{x} = \tilde{x}^+(+)\tilde{x}^-$. We now have the following definition.

Definition 10.8.3 (Fuzzy scalar product). *Let $\tilde{x}, \tilde{y} \in F^n(\mathbb{R})$. Then the fuzzy scalar product of $\tilde{x}$ and $\tilde{y}$, denoted by $\ll \tilde{x}, \tilde{y} \gg$, is defined by*

$$\ll \tilde{x}, \tilde{y} \gg = ((\tilde{x}_1(\cdot)\tilde{y}_1)(+)\ldots(+)(\tilde{x}_n(\cdot)\tilde{y}_n)).$$

We now introduce an appropriate primal-dual pair of fuzzy linear programming problems and establish relevant duality theorems for the same. For this let $\tilde{c} \in F^n(\mathbb{R})$, $\tilde{b} \in F^m(\mathbb{R})$ and $\tilde{A} = (\tilde{a}_{ij})$ be a $(m \times n)$ fuzzy matrix with $\tilde{a}_{ij} \in F(\mathbb{R})$. Let $\tilde{A}_i$ $(i = 1, 2 \ldots, m)$ and $\tilde{A}_j$ $(j = 1, 2 \ldots, n)$ respectively be the ith row and j^{th} column of $\tilde{A}$. We now consider the following problems $(\widetilde{LP})$ and $(\widetilde{LD})$

$(\widetilde{LP})$ min $(\tilde{c}_1(\cdot)\tilde{I}x_1)(+)(\tilde{c}_2(\cdot)\tilde{I}x_2)(+)\ldots(+)(\tilde{c}_n(\cdot)\tilde{I}x_n)$
subject to,

$$(\tilde{a}_{i1}(\cdot)\tilde{I}x_1)(+)\ldots(+)(\tilde{a}_{in}(\cdot)\tilde{I}x_n) \geqq \tilde{b}_i, \ (i = 1, 2 \ldots, m),$$
$$x_1, x_2, \ldots, x_n \geq 0.$$

$(\widetilde{LD})$ max $(\tilde{b}_1(\cdot)\tilde{I}y_1)(+)(\tilde{b}_2(\cdot)\tilde{I}y_2)\ldots(+)(\tilde{b}_n(\cdot)\tilde{I}y_n)$
subject to,

$$(\tilde{a}_{1j}(\cdot)\tilde{I}y_1)(+)\ldots(+)(\tilde{a}_{mj}(\cdot)\tilde{I}y_m) \leqq \tilde{c}_j, \ (j = 1, 2 \ldots, n),$$
$$y_1, y_2, \ldots, y_m \geq 0.$$

In terms of our notations for fuzzy scalar product, problems $(\widetilde{LP})$ and $(\widetilde{LD})$ can also be expressed as

$(\widetilde{LP1})$ min $\ll \tilde{c}, x \gg$
subject to,

$$\tilde{A}x \geqq \tilde{b},$$
$$x \geq 0,$$

and

$(\widetilde{LD1})$ max $\ll \tilde{b}, y \gg$
subject to,

$$\tilde{A}^T y \leqq \tilde{c},$$
$$y \geq 0.$$

Here in the scalar product $\ll \tilde{c}, x \gg$ (respectively $\ll \tilde{b}, y \gg$) x (respectively y) is treated as $\tilde{x}$ (respectively $\tilde{y}$) with membership function of $\tilde{x}$ (respectively $\tilde{y}$) as the characteristic function of x (respectively y).

Definition 10.8.4 (Feasible region of $(\widetilde{LP1})$). *Let* $X = \{ x \in \mathbb{R}^n : \ll \tilde{A}_i, x \gg \geqq \tilde{b}_i,\ (i = 1,2,\ldots,m),\ x \geq 0 \}$ *i.e.* $X = \{x \in \mathbb{R}^n : \tilde{A}x \geqq \tilde{b}, x \geq 0\}$. *Then* X *is called the feasible region of* $(\widetilde{LP1})$.

Definition 10.8.5 (Solution of $(\widetilde{LP1})$). *A point* $x^* \in \mathbb{R}^n$ *is called a solution of the problem* $(\widetilde{LP1})$ *if there does not exist any* $x\ (\neq x^*)$ *such that* $\ll \tilde{c}, x^* \gg \succeq\ \ll \tilde{c}, x \gg$. *In that case the set* $\{\ll \tilde{c}, x^* \gg : x^*$ *is a solution of* $(\widetilde{LP1})\}$ *is denoted by* $Min_P(\tilde{A}, \tilde{b}, \tilde{c})$.

In a similar manner we define the feasible region of $(\widetilde{LD1})$ as $Y = \{y \in \mathbb{R}^m : \ll \tilde{A}_j, y \gg \leqq \tilde{c}_j,\ (j = 1,2\ldots n),\ y \geq 0\}$ i.e. $Y = \{y \in \mathbb{R}^m : \tilde{A}^T y \leqq \tilde{c},\ y \geq 0\}$. Further a point $y^* \in \mathbb{R}^m$ is called a solution of the $(\widetilde{LD1})$ if there does not exist any $y\ (\neq y^*)$ such that $\ll \tilde{b}, y^* \gg \preceq \ll \tilde{b}, y \gg$. In that case the set $\{\ll \tilde{b}, y^* \gg :\ y^*$ is a solution of $(\widetilde{LD1})\}$ is denoted by $Max_D(\tilde{A}, \tilde{b}, \tilde{c})$.

The following lemmas will be useful in the sequel.

Lemma 10.8.1. *Let* $\tilde{A} = (\tilde{a}_{ij})$ *be an* $m \times n$ *fuzzy matrix. Let* $x \in \mathbb{R}^n, x \geq 0$ *and* $y \in \mathbb{R}^m$, $y \geq 0$. *Then* $\ll \tilde{A}^T y, x \gg = \ll y, \tilde{A}x \gg$

Lemma 10.8.2. *Let* $w \in \mathbb{R}^n$ *be nonnegative and* $x \in \mathbf{F}^n(\mathbb{R})$. *Then* $\ll w, \tilde{x} \gg_\alpha^L = < w, x_\alpha^L >$ *and* $\ll w, x \gg_\alpha^R = < w, x_\alpha^R >$.

The proofs of above lemmas follow directly from the definitions of $(+), (\cdot)$ and the fact that the addition and multiplication of closed intervals are both associative and commutative.

Theorem 10.8.1 (Weak duality theorem). *Let* $x \in X$ *and* $y \in Y$. *Then* $\ll \tilde{c}, x \gg \geqq \ll \tilde{b}, y \gg$. *Further* $Min_P(\tilde{A}, \tilde{b}, \tilde{c}) \geqq Max_D(\tilde{A}, \tilde{b}, \tilde{c})$.

Proof. We observe that

$$\begin{aligned} \ll \tilde{c}, x \gg\ &\geqq \ll \tilde{A}^T y, x \gg, \text{ (because } \tilde{A}^T y \leqq \tilde{c}, x \geq 0, y \geq 0), \\ &= \ll y, \tilde{A}x \gg, \text{ (by Lemma 10.8.1)}, \\ &\geqq \ll y, \tilde{b} \gg, \text{ (because } \tilde{A}x \geqq \tilde{b}, x \geq 0, y \geq 0). \end{aligned}$$

Therefore $\ll \tilde{c}, x \gg \geqq \ll \tilde{b}, y \gg$. As this relation holds for all feasible solution y of $(\widetilde{LD1})$, it implies that $\ll \tilde{c}, x \gg \geqq Max_D(\tilde{A}, \tilde{b}, \tilde{c})$ for all feasible solution x of $(\widetilde{LP1})$. This gives $Min_P(\tilde{A}, \tilde{b}, \tilde{c}) \geqq Max_D(\tilde{A}, \tilde{b}, \tilde{c})$. Here we are using the ordering between two subsets of fuzzy numbers as per Definition 10.8.2

Corollary 10.8.1 *Let $x^* \in X, y^* \in Y$ and $\ll \tilde{c}, x^* \gg = \ll \tilde{b}, y^* \gg$. Then $\ll \tilde{c}, x^* \gg \leqq \ll \tilde{c}, x \gg$ for all $x \in X$ and $\ll \tilde{b}, y^* \gg \geqq \ll \tilde{b}, y \gg$ for all $y \in Y$.*

Proof. Let y be an feasible solution of $(\widetilde{LP1})$. Then by the weak duality theorem (Theorem 10.8.1) we have $\ll \tilde{c}, x^* \gg \geqq \ll \tilde{b}, y \gg$ i.e. $\ll \tilde{b}, y^* \gg \geqq \ll \tilde{b}, y \gg$. The other part of the corollary is analogous.

If we agree to denote the objective functions of $(\widetilde{LP1})$ and $(\widetilde{LD1})$ by fuzzy numbers $\tilde{P}(x)$ and $\tilde{D}(x)$ and the corresponding α-cuts by $[P^L_\alpha, P^R_\alpha]$ and $[D^L_\alpha, D^R_\alpha]$ respectively, then the statement $\ll \tilde{c}, x^* \gg \leqq \ll \tilde{c}, x \gg$ for all $x \in X$ means that for all $\alpha \in [0,1]$. $P^L_\alpha \leq D^L_\alpha$ and $P^R_\alpha \leq D^R_\alpha$, i.e. $\nexists \alpha \in [0,1]$ such that $P^L_\alpha > D^L_\alpha$ and $P^R_\alpha > D^R_\alpha$. Therefore x^* is certainly a solution of $(\widetilde{LP1})$ and therefore $\ll \tilde{c}, x^* \gg \in Min_P(\tilde{A}, \tilde{b}, \tilde{c})$. But $\ll \tilde{c}, x^* \gg \leqq \ll \tilde{c}, x \gg$ for all $x \in X$ is more than just saying that x^* is a solution of $(\widetilde{LP1})$. Similar statements hold for $(\widetilde{LD1})$ as well.

Definition 10.8.6 (No duality gap property). *The pair $(\widetilde{LP1})$ and $(\widetilde{LD1})$ is said to have no duality gap property (Wu [82]) if*

$$Min_P(\tilde{A}, \tilde{b}, \tilde{c}) \cap Max_D(\tilde{A}, \tilde{b}, \tilde{c}) \neq \phi \text{ (empty set).}$$

Here though the sets $Min_P(\tilde{A}, \tilde{b}, \tilde{c})$ and $Max_D(\tilde{A}, \tilde{b}, \tilde{c})$ are fuzzy sets, the above intersection not being empty is to be understood in the sense that there is a fuzzy element common to both of these fuzzy sets.

Lemma 10.8.3. *Let the problems $(\widetilde{LP1})$ and $(\widetilde{LD1})$ have no duality gap. Then there exist $x^* \in X, y^* \in Y$ such that $\ll \tilde{c}, x^* \gg \leqq \ll \tilde{c}, x \gg$ for all $x \in X$ and $\ll \tilde{b}, y^* \gg \geqq \ll \tilde{b}, y \gg$ for all $y \in Y$.*

Proof. Since $Min_P(\tilde{A}, \tilde{b}, \tilde{c}) \cap Max_D(\tilde{A}, \tilde{b}, \tilde{c}) \neq \phi$, there exists $x^* \in X$, $y^* \in Y$ such that $\ll \tilde{c}, x^* \gg \in Min_P(\tilde{A}, \tilde{b}, \tilde{c})$, $\ll \tilde{b}, y^* \gg \in Max_D(\tilde{A}, \tilde{b}, \tilde{c})$ and $\ll \tilde{c}, x^* \gg = \ll \tilde{b}, y^* \gg$. The rest of the proof now follows from the Corollary 10.8.1.

The "no duality gap property" needed to establish Lemma 10.8.3 seems to be a very strong requirement almost equivalent to the result itself. However the below given theorem shows that to establish the strong duality theorem, the requirement of "no duality gap property" can be somewhat relaxed to the requirement that the sets $Arg\text{-}Min_P(\tilde{A}, \tilde{b}, \tilde{c})$ and $Arg\text{-}Max_D(\tilde{A}, \tilde{b}, \tilde{c})$ are nonempty.

We now introduce the sets $Arg\text{-}Min_P(\tilde{A}, \tilde{b}, \tilde{c})$ and $Arg\text{-}Max_D(\tilde{A}, \tilde{b}, \tilde{c})$. For this let A^L_α (respectively A^R_α) be the matrix whose $(i,j)^{th}$ entry is

$(a_{ij})^L_\alpha$ (respectively $(a_{ij})^R_\alpha$). Now corresponding to the problem $(\widetilde{LP1})$ we construct the following two α-level (crisp) linear programming problems for each $\alpha \in [0,1]$

$(LP1)^L_\alpha$ min $\ll c^L_\alpha, x \gg$
subject to,

$$A^L_\alpha x \geq b^L_\alpha,$$
$$x \geq 0,$$

and

$(LP1)^R_\alpha$ min $\ll c^R_\alpha, x \gg$
subject to,

$$A^R_\alpha x \geq b^R_\alpha,$$
$$x \geq 0.$$

In a similar manner we construct α-level (crisp) linear programming problems corresponding to $(\widetilde{LD1})$. These are

$(LD1)^L_\alpha$ max $\ll b^L_\alpha, y \gg$
subject to,

$$(A^L_\alpha)^T y \leq c^L_\alpha,$$
$$y \geq 0.$$

and

$(LD1)^R_\alpha$ max $\ll b^R_\alpha, y \gg$
subject to,

$$(A^R_\alpha)^T y \leq c^R_\alpha,$$
$$y \geq 0.$$

Here it is to be noted that $(LD1)^L_\alpha$ is the dual of $(LP1)^L_\alpha$ and similarly $(LD1)^R_\alpha$ is the dual of $(LP1)^R_\alpha$. Now we define

$$(Arg\text{-}Min_P(\tilde{A},\tilde{b},\tilde{c}))^L_\alpha = \text{set of all finite optimal solutions of } (LP1)^L_\alpha,$$
$$(Arg\text{-}Min_P(\tilde{A},\tilde{b},\tilde{c}))^R_\alpha = \text{set of all finite optimal solutions of } (LP1)^R_\alpha,$$
$$(Arg\text{-}Min_P(\tilde{A},\tilde{b},\tilde{c})^L = \bigcap_{0\leq\alpha\leq 1} (Arg\text{-}Min_P(\tilde{A},\tilde{b},\tilde{c}))^L_\alpha,$$
$$(Arg\text{-}Min_P(\tilde{A},\tilde{b},\tilde{c})^R = \bigcap_{0\leq\alpha\leq 1} (Arg\text{-}Min_P(\tilde{A},\tilde{b},\tilde{c}))^R_\alpha,$$
$$Arg\text{-}Min_P(\tilde{A},\tilde{b},\tilde{c}) = (Arg\text{-}Min_P(\tilde{A},\tilde{b},\tilde{c}))^L \cap (Arg\text{-}Min_P(\tilde{A},\tilde{b},\tilde{c}))^R.$$

The set $Arg\text{-}Max_D(\tilde{A},\tilde{b},\tilde{c})$ is defined analogously by using the (crisp) problems $(LD1)^L_\alpha$ and $(LD1)^R_\alpha$. Also $x^* \in Arg\text{-}Min_P(\tilde{A},\tilde{b},\tilde{c})$ means that

$\ll \tilde{c}, x^* \gg \in Min_P(\tilde{A}, \tilde{b}, \tilde{c})$. Similarly for y^* we have $y^* \in Arg\text{-}Max_D(\tilde{A}, \tilde{b}, \tilde{c})$ gives $\ll \tilde{b}, y^* \gg \in Max_D(\tilde{A}, \tilde{b}, \tilde{c})$.

Theorem 10.8.2 (Strong duality theorem). *Let the sets* $Arg\text{-}Min_P(\tilde{A}, \tilde{b}, \tilde{c})$ *and* $Arg\text{-}Max_D(\tilde{A}, \tilde{b}, \tilde{c})$ *be nonempty. Then the problems* $(\widetilde{LP}1)$ *and* $(\widetilde{LD}1)$ *have no duality gap.*

Proof. Let $x^* \in Arg\text{-}Min_P(\tilde{A}, \tilde{b}, \tilde{c})$ and $y^* \in Arg\text{-}Max_D(\tilde{A}, \tilde{b}, \tilde{c})$. Then by the crisp linear programming duality, $< c^L_\alpha, x^* > = < b^L_\alpha, y^* >$ and $< c^R_\alpha, x^* > = < b^R_\alpha, y^* >$ for all $\alpha \in [0, 1]$. Since x^* and y^* are nonnegative, these relations by Lemma 10.8.2, give $\ll \tilde{c}, x^* \gg = \ll \tilde{b}, y^* \gg$, which because of Lemma 10.8.3 proves the theorem.

The next question is how to check that the sets $Arg\text{-}Min_P(\tilde{A}, \tilde{b}, \tilde{c})$ and $Arg\text{-}Max_D(\tilde{A}, \tilde{b}, \tilde{c})$ are nonempty. For this Wu [82] gave a sufficient condition which is based on the following definition:

Definition 10.8.7 (Finite intersection property). *A family* $\{K_\alpha\}$ *of sets in a topological space* Ω *is said to have the finite intersection property if every finite sub family of* $\{K_\alpha\}$ *has a nonempty intersection.*

Theorem 10.8.3 *If the families* $\left\{\left(Arg\text{-}Min_P(\tilde{A}, \tilde{b}, \tilde{c})^L_\alpha\right) \cap \left(Arg\text{-}Min_P(\tilde{A}, \tilde{b}, \tilde{c})^R_\alpha\right) : \alpha \in [0, 1]\right\}$ *and* $\left\{\left(Arg\text{-}Max_D(\tilde{A}, \tilde{b}, \tilde{c})^L_\alpha\right) \cap \left(Arg\text{-}Max_D(\tilde{A}, \tilde{b}, \tilde{c})^R_\alpha\right) : \alpha \in [0, 1]\right\}$ *have the finite intersection property then problem* $(\widetilde{LP}1)$ *and* $(\widetilde{LD}1)$ *have no duality gap.*

We shall not prove the above theorem here and shall refer to Wu [82] in this connection.

Remark 10.8.4. A close look at the discussion on strong duality theorem for the pair $(\widetilde{LP}1)$ and $(\widetilde{LD}1)$ suggests that the status is far from satisfactory. Firstly there seems to be no simple way to check if the stated families in Theorem 10.8.3 have the finite intersection property so that the "no duality gap property" can be guaranteed. Further unlike the crisp case where existence of optimal solution to the primal guarantees the existence of optimal solution to the dual, here both primal and dual $(\widetilde{LP}1)$ and $(\widetilde{LD}1)$ are assumed to have solutions. Also there seems to be no known general class of fuzzy linear programming problems, even with TFN data , for which these duality results are known to hold. This puts a major limitation on this approach.

10.9 Conclusion

In this chapter we have included comparatively newer results on fuzzy linear programming. These results are representative of the general direction in which the current research in the area of fuzzy linear programming is progressing. Modality constrained programming has already established its importance in the area of fuzzy decision making but the same can not possibly be said about duality in fuzzy linear programming at this stage. However results are available on duality in fuzzy nonlinear programming with fuzzy coefficients e.g. Wu ([81], [82], [83]) and its ramifications with generalized convexity, e.g. Ramik and Vlach [65]. Also there are some other papers on fuzzy optimization problems, e.g. Wu([84], [85] and [86]) which use the fact that the set of all fuzzy numbers can be embedded into a suitable Banach space so that the fuzzy optimization problem can be transformed into a biobjective programming problem. Employing this approach Wu [83] derived the Karush-Kuhn-Tucker optimality conditions for the fuzzy optimization problem with fuzzy coefficients and also obtained some computational procedures for the same.

References

1. J.M. Adamo, Fuzzy decision trees, *Fuzzy Sets and Systems*, Vol. 4, pp. 207-219, 1980.
2. M.S. Bazaraa, H.D. Sherali and C.M. Shetty, *Nonlinear Programming: Theory and Algorithms*, 2nd Edition, John Wiley and Sons, Inc., New York, NY, 1993. 2nd Edition, John Wiley and Sons, Inc., New York, NY, 1990.
3. E.M.L. Beale, On quadratic programming, *Naval Research Logistics Quarterly*, Vol. 6, pp. 227-244, 1959.
4. C.R. Bector and S. Chandra, On duality in linear programming under fuzzy environment, *Fuzzy Sets and Systems*, Vol. 125, pp. 317-325, 2000.
5. C.R. Bector, S. Chandra and M. Singh, Quadratic programming under fuzzy environment, *Research Report*, Faculty of Management, University of Manitoba, Canada, 2002.
6. C.R. Bector and S. Chandra and V. Vijay, Matrix games with fuzzy goals and fuzzy linear programming duality, *Fuzzy Optimization and Decision Making*,Vol. 3, pp. 263-277, 2004.
7. C.R.Bector and S.Chandra and V. Vijay, Duality in linear programming with fuzzy parameters and matrix games with fuzzy pay-offs, *Fuzzy Sets and Systems*, Vol. 146, pp. 253-269, 2004.
8. R. Bellman and M. Giertz, On the analytic formulation of the theory of fuzzy sets, *Information Sciences*, Vol. 5, pp. 149-156, 1973.
9. R.E. Bellman, and L.A. Zadeh, Decision making in a fuzzy environment, *Management Science*, Vol. 17, pp. 141-164, 1970.
10. L. Campos, Fuzzy linear programming models to solve fuzzy matrix games, *Fuzzy Sets and Systems*, Vol. 32, pp. 275-289, 1989.
11. C. Carlsson and Fullér, R. , *Fuzzy Reasoning in Decision Making and Optimization*, Physica-Verlag, 2002.
12. S. Chanas, The use of parametric programming in fuzzy linear programming problems, *Fuzzy sets and Systems*, Vol. 11, pp.243-251, 1983.
13. A. Charnes, Constrained games and linear programming, *Proc. Nat. Acad. Sci.*, Vol. 39, pp. 639-641, 1953.
14. D. Dubois and H. Prade, Fuzzy numbers: an overview, in J.C. Bezdek (ed.) *Analysis of Fuzzy Information*, Mathematics and Logic, CRC Press, Boca Raton, FL, 1997.

15. D. Dubois and H. Prade, *Fuzzy Sets and Systems: Theory and Applications*, Academic Press, New York, 1980.
16. D. Dubois and H. Prade, Ranking fuzzy numbers in the setting of possibility theory, *Information Sciences*, Vol. 30, pp. 183-224, 1980.
17. D. Dumitrescu, L. Lazzzerini, and L.C. Jain, *Fuzzy Sets and Their Application to Clustering and Training*, CRC Press LLC, Florida (USA), 2000.
18. H.F. Wang and C.F. Fu, A generalization of fuzzy programming with preemptive structure, *Computational Operational Research*, Vol. 24, pp. 819-248, 1997.
19. N. Furukawa, A parametric total order on fuzzy numbers and a fuzzy shortest route problem, *Optimization*, Vol. 30, pp. 367-377, 1994.
20. S.M. Guu and Y.K. Wu, Weighted coefficients in two- phase approach for solving the multiple objective programming problems, *Fuzzy sets and Systems*, Vol. 85, pp.45-48, 1997.
21. S.M. Guu and Y.K. Wu, Two phase approach for solving the fuzzy linear programming problems, *Fuzzy sets and Systems*, Vol. 107, pp.191-195, 1999.
22. H. Hamacher, H. Leberling and H.-J. Zimmermann, Sensitivity analysis in fuzzy linear programming, *Fuzzy Sets and Systems* , Vol. 1, pp. 269-281, 1978.
23. E. L. Hannan, On fuzzy goal programming, *Decision Sciences*, Vol. 12, pp. 522-531, 1981.
24. E. L. Hannan, Linear programming with multiple fuzzy goals, *Fuzzy Sets and Systems*, Vol. 6, pp. 819-829, 1981.
25. M. Inuiguchi, H.Ichihashi and Y. Kume, A solution algorithm for fuzzy linear programming with piece wise membership functions, *Fuzzy Sets and Systems*, Vol. 34, pp. 15-31, 1990.
26. M. Inuiguchi, H.Ichihashi and Y. Kume, Relationship between modality constrained programming problems and various fuzzy mathematical programming problems", *Fuzzy sets and Systems*, Vol. 49, pp. 243-259, 1992.
27. M. Inuiguchi, H.Ichihashi and Y. Kume, Some properties of extended fuzzy preference relations using modalities", *Information Sciences*, Vol. 61, pp. 187-209, 1992.
28. M. Inuiguchi, H.Ichihashi and Y. Kume, Modality constrained programming problems: a unified approach to fuzzy mathematical programming problems in the setting of possibility theory, *Information Sciences*, Vol. 67, pp. 93-126, 1993.
29. M. Inuiguchi, J. Ramik, T. Tanino and M. Vlach, Satisficing solutions and duality in interval and fuzzy linear programming, *Fuzzy Sets and Systems*, Article in Press.
30. W. Jianhua, *The Theory of Games*, Tsinghua University Pres, Beijing Clarendon Press, Oxford, 1988.
31. S. Karlin, *Mathematical Methods and Theory in Games, Programming and Economics*, Addison-Wesley Publishing Company, Inc., USA, 1959.
32. A. Kaufmann and M.M. Gupta, *Fuzzy Mathematical Models in Engineering and Management Science*, North Holland, Amsterdam, 1988.
33. A. Kaufmann and M.M. Gupta, *Introduction to Fuzzy Arithmetic: Theory and Applications*, Van Norstrand Reinhold, Newyork, 1991.
34. T. Kawaguchi and Y. Maruyama, A note on minmax (maxmin) programming, *Management Science*, Vol. 22, pp. 670-676, 1976.
35. G. J. Klir and B. Yuan, *Fuzzy Sets and Fuzzy Logic: Theory and Applications*, Prentice-Hall, Inc., N.J. (USA), 1996.
36. J.J. Lai and C.L. Hwang, Possibilistic linear programming for managing interest rate risk, *Fuzzy Sets and Systems*, Vol. 49, pp. 121-133, 1992.

37. Y.J. Lai and C.L. Hwang, *Fuzzy Mathematical Programming*, Springer, Verlag, Heidelberg, 1992.
38. Y.J. Lai and C.L. Hwang, *Fuzzy Multiple Objective Decision Making: Methods and Application*, Lecture Notes in Economics and Mathematical Systems, Vol. 404, Springer, Verlag, Heidelberg, 1994.
39. D.-F. Li, A fuzzy multi-objective approach to solve fuzzy matrix games, *The Journal of Fuzzy mathematics*, Vol. 7, pp.907-912, 1999.
40. D.-F. Li, Fuzzy constrained matrix games with fuzzy payoffs, *The Journal of Fuzzy Mathematics*, Vol. 7, 873-880, 1999.
41. D.-F. Li and C.-T. Cheng, Fuzzy multiobjective programming methods for fuzzy constrained matrix games with fuzzy numbers, *International Journal of Uncertainty, Fuzziness and Knowledge-Based Systems*, Vol. 10(4), 2002.
42. D.-F. Li and J.B.Yang, Two level linear programming approach to solve fuzzy matrix games with fuzzy pay offs, Manchester School of Management, University of Manchester Institute of Science and Technology, UK, Unpublished preprint, 2004.
43. H.L. Li and C.S. Yu, A fuzzy multiobjective program with quasiconcave membership functions and fuzzy coefficients, *Fuzzy Sets and Systems*, Vol. 109, pp. 59-81, 2000.
44. R.J. Li, *Multiple Objective Decision Making in Fuzzy Environment*, Ph.D. Dissertation, 1990, Department of Industrial Engineering, Kansas State University, Manhattan, KS.
45. C.C. Lin, A weighted max-min model for fuzzy goal programming, *Fuzzy Sets and Systems*, Vol. 142, pp. 407-420, 2004.
46. E.S.-Lee and R.J-Li, Fuzzy multiple objective programming and compromise programming with pareto optimum, *Fuzzy Sets and Systems*, Vol. 53, pp. 275-288, 1993.
47. C.C. Lin and A.P. Chen, A generalization of Yang et al.'s method for fuzzy programming with piecewise linear membership functions, *Fuzzy Sets and Systems*, Vol. 132, pp. 347-352, 2002.
48. C. T. Lin and C. S. G. Lee, *Neural Fuzzy Systems: A Neuro-Fuzzy Synergism to Intelligent Systems*, Prentice-hall International, Inc, N.J. (USA), 1996.
49. T. Maeda, Characterization of equilibrium strategy of the bi-matrix game with fuzzy payoffs, *Journal of Mathematical Analysis and Applications*, Vol. 251, pp. 885-896, 2000.
50. T. Maeda, On characterization of equilibrium strategy of two person zero sum games with fuzzy pay-offs, *Fuzzy Sets and Systems*, Vol. 139, pp. 283-296, 2003.
51. J.M. Mandel, *Uncertain Rule-Based Fuzzy Logic Systems: Introduction and New Directions*, Prentice Hall Inc., 2001.
52. O.L. Mangasarian and H. Stone, Two person non zero sum games and quadratic programming, *Journal of Mathematical Analysis and Applications*, Vol. 9, pp. 348-355, 1964.
53. O.L. Mangasarian, *Nonlinear Programming*, SIAM, Philadelphia, PA, 1994.
54. R.H. Mohamed, The relationship between goal programming and fuzzy programming, *Fuzzy Sets and Systems*, Vol. 89, pp. 215-222, 1997.
55. R. E. Moore, *Interval Analysis*, Englewood Cliffs, Prentice Hall, N.J., 1966.
56. R. E. Moore, *Methods and Applications of Interval Analysis*, SIAM, Philadelphia.
57. K. Nakamura, Some extension of fuzzy linear programming, *Fuzzy Sets and Systems*, Vol. 14, pp. 211-229, 1984.

58. R. Narasimhan, Goal programming in a fuzzy environment, *Decision Sciences*, Vol. 11, pp. 325-336, 1980.
59. R. Narasimhan, On fuzzy goal programming-some comments, *Decision Sciences* Vol. 12, pp. 532-538, 1981.
60. J.F. Nash, Non cooperative Games, *Annals of Mathematics*, Vol. 54, pp. 286-295, 1951.
61. I. Nishizaki and M. Sakawa, *Fuzzy and Multiobjective Games for Conflict Resolution*, Physica-verleg, Heidelberg, 2001.
62. G. Owen, *Game Theory*, Academic Press, San Diego, 1995.
63. B.B. Pal, B.N. Moitran and U. Maulik, A goal programming procedure for fuzzy multiobjective linear fractional programming problem, Vol.139, pp. 395-405, 2003.
64. T. Parthasarathy and T.E.S. Raghavan, *Some Topics in Two Person Games*, American Elsevier Publishing Company, Inc., New York (USA),1971.
65. J. Ramik and M. Vlach, *Generalized Convexity in Fuzzy Optimization and Decision Analysis*, Kluwer Academic Publishers, 2001.
66. J. Ramik and J. Rimanek, Inequality relation between fuzzy numbers and its use in fuzzy optimization, *Fuzzy Sets and Systems*, Vol. 16, pp. 123-138, 1985.
67. W. Rödder and H.-J. Zimmermann. *Duality in fuzzy linear programming, External Methods and System Analysis*, in A.V. Fiacco and K.O. Kortanek (eds.), Springer-Verleg, Heidelberg, 1980.
68. M. Sakawa and I. Nishizaki, Maxmin solutions for fuzzy multiobjective matrix games, *Fuzzy Sets and Systems*, Vol. 61, pp. 265-275, 1994.
69. B. Schweizer and A. Sklar, Statistical metric spaces, *Pacific Journal of Mathematics*, Vol. 10, pp. 313-334, 1960.
70. I.M. Stancu-Minasian, *Stochastic Programming with Multiple Objective Functions*, D.Reidal Publishing Company, Dordrecht, Holland, 1984.
71. R.E. Steuer, *Multi Criteria Optimization Theory, Computation and Application*, John Wiley and Sons, Inc., 1986.
72. Taha, H.A, *Operational Research : An introduction*, Macmillan, 3rd Edition, 1982.
73. C. Van De Panne, Programming with a quadratic constraint, *Management Sciences*, Vol.12, pp. 798-815, 1966.
74. J.L. Verdegay, A dual approach to solve the fuzzy linear programming problem, *Fuzzy Sets and Systems*, Vol. 14, pp. 131-141, 1984.
75. J.L. Verdegay, Fuzzy mathematical programming, in M.M. Gupta and E. Sanchez (eds.), *Fuzzy Information and Decision Processes*, North Holland, Amsterdam, 1982.
76. V. Vijay, Constrained matrix games with fuzzy goals, Conference on *Fuzzy Logic and its Application in Technology and Management*, Indian Institute of Technology, Kharagpur, India, 2004.
77. V. Vijay, S.Chandra and C.R. Bector, Bi-matrix games with fuzzy goals and fuzzy pay-offs, *Fuzzy Optimization and Decision Making*, To appear in Dec. 2004.
78. H.F. Wang and C.C. Fu, A generalization of fuzzy programming with preemptive structure, *Computational Operational Research*, Vol. 24, pp. 819-828, 1997.
79. B. Werners, Interactive multiple objective programming subject to flexible constraints, *European Journal of Operations research*, Vol. 31, pp. 342-349, 1987.
80. P. Wolfe, The simplex method for quadratic programming, *Econometrica*, Vol. 27, pp. 382-398, 1959.

81. H.C. Wu, Duality theorems in fuzzy mathematical programming problems based on the concept of necessity, *Fuzzy Sets and Systems*, Vol. 139, pp. 363-377, 2003.
82. H.-C. Wu, Duality theory in fuzzy linear programming problems with fuzzy coefficients, *Fuzzy Optimization and Decision Making*, Vol. 2, pp. 61-73, 2003.
83. H.C. Wu, Saddle point optimality conditions in fuzzy optimization, *Fuzzy Optimization and Decision Making*, Vol. 2, pp. 261-273, 2003.
84. H.C. Wu, Evaluate fuzzy optimization problems based on bi-objective programming problems, *Computers and Mathematics with Applications*, Vol. 47, pp. 1263-1271, 2004.
85. H.C. Wu, Fuzzy optimization problems based on the embedding theorem and possibility and necessity measures, *Mathematical and Computer Modelling*, Article in Press, 2003.
86. H.C. Wu, An (α, β)-optimal solution concept in fuzzy optimization problems, *Optimization*, Vol.53, pp. 203-221, 2004.
87. R.R. Yager, A procedure for ordering fuzzy numbers of the unit interval, *Information Sciences*, Vol. 24, pp. 143-161, 1981.
88. T. Yang, J.P. Ignizio and H.J. Kim, Fuzzy programming with nonlinear membership functions: piecewise linear approximation, *Fuzzy Sets and Systems*, Vol.41, pp. 39-53, 1991.
89. L. A. Zadeh, Fuzzy Sets, *Information and Control*, Vol. 8, pp. 338-353, 1965.
90. H.-J. Zimmermann, Fuzzy programming and linear programming with several objective functions, *Fuzzy Sets and Systems*, Vol. 1, pp. 45-55, 1978.
91. H. -J. Zimmermann, *Fuzzy Set Theory and Its Applications*, 3rd Edition, Kluwer Academic Publishers, Nowell, MA, USA, 1996.

Index